GIS FOR HEALTH *and the* ENVIRONMENT

Proceedings of an International Workshop
held in Colombo, Sri Lanka
5 – 10 September 1994

EDITED BY
Don de Savigny and Pandu Wijeyaratne

INTERNATIONAL DEVELOPMENT RESEARCH CENTRE
Ottawa • Cairo • Dakar • Johannesburg • Montevideo
Nairobi • New Delhi • Singapore

Published by the International Development Research Centre
PO Box 8500, Ottawa, ON, Canada K1G 3H9

June 1995

de Savigny, D.
Wijeyaratne, P.
IDRC, Ottawa, ON CA

GIS for health and the environment : proceedings of an international workshop held in Colombo, Sri Lanka, 5–10 September 1994. Ottawa, ON, IDRC, 1995. 184 p. : ill.

/Geographic information systems/, /public health/, /environmental health/, /disease control/, /developing countries/ — /remote sensing/, /spatial analysis/, /disease vectors/, /disease transmission/, /tropical diseases/, /case studies/, /conference reports/, /annotated bibliographies/, references.

UDC: 007:910:614 ISBN: 0-88936-766-3

A microfiche edition is available.

IDRC's Focus Series presents research results and scholarly studies from the Centre's six core themes of program support.

Contents

Appendicies

Foreword[1]

Don de Savigny,[2a] Luc Loslier,[b] and Jim Chauvin[c]

Canada's unique geography and population distribution has prompted innovation in such areas as natural resource management and regional planning. One such innovation was the Geographic Information System (GIS), a concept that originated in Canada three decades ago. This concept is now being applied by the health sector as the demand increases for information and analysis on the relationship between people, their environment and their health. GIS can provide health researchers, planners, program managers, and policymakers with novel information about the distribution and interaction of disease risk factors, patterns of morbidity and mortality, and the allocation of health resources. Now that many commercially available GIS software packages are becoming increasingly user friendly, and can be run on PC computers, this important tool is being actively explored for use in health improvements in developing countries.

So what is a GIS? Essentially, it is a computer-assisted information management system of geographically-referenced data. It contains two closely integrated databases: one spatial (locational) and the other attribute (statistical). The spatial database contains information in the form of digital coordinates, usually from maps or from remote-sensing. These can be points (for example, health clinics or houses), lines (rivers and roads), or polygons (district health units). The attribute database contains information about the characteristics or qualities of the spatial features, for example, demographic information, immunization rates, number of health personnel at a clinic, or the type of road access.

A GIS differs from conventional computer-assisted mapping and statistical analysis systems. Although computer-assisted cartographic systems emphasize map production and presentation of spatial data, they cannot analyze spatially-defined statistical data. Statistical analysis systems, on the other hand, analyze aspatial data. A GIS blends these into a more powerful analytical tool. The system can be applied to a variety of problems. For example, a GIS can be

---

[1]This text is based on an article published by the authors in Synergy — Canadian Initiatives for International Health 1994, 6(4), 1–2.

[2a]Health Sciences Division, International Development Research Centre, Ottawa, Canada.

[b]Department of Geography, Université du Québec à Montréal, Canada.

[c]International Health Program, Canadian Public Health Association, Ottawa, Canada.

used to investigate questions about location (what are the attributes at a specific place?), condition (where are the sites which possess certain attributes?), trends (how do attributes change spatially over time?), routing (what is the shortest/least expensive/most cost-effective path between places?), and patterns (what is the distribution of attributes and the process/reason accounting for their distribution?). A GIS can also be used to simulate "what if" scenarios (modelling).

For example, let's say someone wanted to identify districts that have a health centre and where the immunization coverage rate for the third dose of DPT (diphtheria, pertussis, and tetanus) vaccine for children 0–1 years of age is less than 50%. We also want to compare this information to data about family size, the availability of health personnel, traditional water supply sources (lakes, rivers, ponds), and improved water supply facilities.

A GIS is required because the health service facility, immunization, demographic, health personnel, hydrological system, and improved water supply facility databases have different geographically-defined information. The GIS can also define very specific correlations. For example, by combining areas where the immunization coverage rate is low, the number of young children per family is high and the access to improved water supply sources is difficult, the analyst or researcher can define areas and populations at greater or lesser risk.

Traditional database analysis systems are only adequate for analysis of attributes belonging to the same entities. But the analytical tools available within a GIS make it possible to "overlay" the databases for different entities. This is similar to overlaying transparent maps for each database on top of each other. The maps produced by the GIS show the correlation among the different databases. An additional comparative advantage of a GIS is its ability to use data from, and interface with, other sophisticated statistical databases.

GIS is a fairly new analytical and planning tool for both the North and the South. Its proponents highlight its capacity to produce a comprehensive and timely analysis of complex databases and its potential to improve data collection, analysis and presentation processes. The visual impact of GIS-produced maps on decision-making and management is a tangible benefit that is often underestimated.

Although the sectors concerned with agriculture, natural resources, urban and regional planning, and tourism in developing countries have been using GIS for many years, the health sector has only recently begun to work with this tool. The papers collected in these proceedings represent examples of some of the first attempts to identify and explore opportunities to apply GIS for health in developing countries.

Preface

Don de Savigny,[1a] Lori Jones–Arsenault,[a] and Pandu Wijeyaratne[b]

IDRC has been supporting the development and application of Geographic Information Systems (GIS) in developing countries since 1986. During an internal review of progress in this field conducted in 1991, we were struck by the fact that despite rapid and productive adoption of this tool by sectors such as agriculture, natural resources, demography, urban and regional planning, and so on, the health sector had not yet begun to explore the potential utility of GIS for either health research or for health programing. A review of the literature at that time revealed only one publication that described a developing-country health application of GIS. Thus, although GIS itself is not new to most developing countries, its extension into the health sector for interdisciplinary health research and health development provides a new and exciting focus.

At this same point in time, both commercial and public domain GIS software are becoming increasingly simplified, affordable, and available on PC computer platforms. We felt that the moment had come to encourage some experiments in the use of GIS in a variety of developing country settings to generate some experiences from which decisions could be taken on the potential utility of GIS to either health research or health development. Since then, the Health Sciences Division of IDRC has helped develop and support 10 such studies. It was always intended that at a critical juncture in this effort we would convene an international workshop of the principal investigators to share their experiences, problems, needs, and visions for the future.

This moment came in September 1994, when the first international workshop on the use of GIS in health in developing countries was convened by IDRC in Colombo, with superb facilities and organization provided by the University of Colombo, Sri Lanka. From these modest beginnings, this workshop attracted much attention, testifying to the rapid uptake of this tool now underway in the South. Papers were presented by all IDRC GIS health projects and a number of other projects currently underway. Forty-five participants from 19

[1a]Health Sciences Division, International Development Research Centre, Ottawa, Canada.

[b]Tropical Disease Prevention, Environmental Health Project, USAID, Arlington, VA, USA (formerly with the Health Sciences Division of the International Development Research Centre).

countries attended, representing a truly interdisciplinary mixture of health, social, environmental, and computer scientists from ministries of health, ministries of environment, nongovernmental organizations, academia, and the private sector. In particular, the workshop focused on the use of GIS as it related to the environmental determinants of health, and the management or minimization of the environment's human impacts.

The objectives of the Workshop were to: explore GIS utility for health surveillance and monitoring; explore GIS utility for health forecasting and control; introduce new or emerging GIS support methodologies; and facilitate networking for health–GIS. The expected outcomes were: enhanced practical ability and understanding of the use, potential, and limitations of GIS for the health sector; enhanced personal and institutional linkages, both South–South and South–North; and dissemination of the current experiences with health based GIS in the South through an appropriate publication. These objectives were predicated on the question of whether GIS has a role in the struggle for health development. We have learned to be wary of technology driven exercises. GIS is usually introduced as a turn-key operation. To be successful, GIS applications in developing countries are being developed, modified, and controlled by people in the South who are in the best position to understand the contexts (social, economic, and political) in which they are implemented and the technical possibilities.

Therefore, the program was in essence dedicated to exploring the state of the art of the GIS – Health nexus, and included general review papers, specific project application papers, extensive discussion sessions, software demonstrations, digitizing and data analysis workshops, question and answer resource sessions, and a needs assessment. The various sessions were facilitated by 20 dedicated computers with a full range of GIS software, and large screen projection facilities for interactive demonstrations.

A selection of the papers presented in Colombo have been edited and collated in these Proceedings. They are organized from the general to the specific. By way of introduction, the Proceedings include a general overview of GIS in the Preface, followed by a detailed introduction to GIS (Steve Reader), a primer on GIS specific to health applications (Luc Loslier), and a perspective on the epidemiological analytical capabilities of GIS (Flavio Nobre). This is followed by a variety of case study papers focusing on tropical disease management in general, specific vector borne diseases in particular, other environment related health problems, and finally, health system applications. The appendix contains an annotated "bibliography" of currently available software, hardware, and resources associated with GIS applications pertinent to developing countries.

In addition to these Proceedings there were a number of other beneficial outcomes of the Colombo Workshop. Most important was the fact that considerable networking occurred and continues today. IDRC recognizes that, at the beginning of any new research endeavour, there are often a number of researchers working in complete isolation. At the time of the Workshop, there were only five publications on GIS health applications in developing countries, yet dozens of researchers were already skilled and active in the field. The workshop provided a powerful opportunity for collaboration, mutual support, and trouble shooting to begin in a way that will serve to accelerate progress significantly. Part of the momentum of Colombo was carried forward to add a health session at the Africa GIS Conference held in Abidjan in March 1995.

The Workshop highlighted several areas of need. Among them: GIS is weak in temporal analysis, particularly spatial-temporal patterns which would be important for incidence modelling; there is a need (and an opportunity) to improve multisectoral approaches via GIS; there is a need for better interfaces with GIS. With regard to the latter, IDRC has identified this as a focus. Beyond direct support to GIS research applications, the Centre has been moving to supporting development of products like Redatam+ and expert decision support systems that interface into or out from GIS respectively.

Acknowledgments

This workshop was locally arranged and hosted by the Malaria Group of the Faculty of Medicine, University of Colombo, Sri Lanka, with professional assistance from Conventions Colombo. Grateful thanks are conveyed to them for the efficient and excellent arrangements of participant travel, conference agenda, and logistics. In addition, the various social activities arranged by the group made this 6-day workshop one of the most efficient scientific meetings ever carried out in a developing country during our tenure with IDRC. Special mention in this respect must be made of Dr Renu Goonewardena and Dr Rajitha Wickremasinghe, and of the overall guidance of Professor Kamini Mendis, together with their dedicated secretaries who worked tirelessly.

Diane Dupuis of the Health Sciences Division, IDRC, worked through the immense task of maintaining close contact with all conference participants — no easy task — and ensuring the receipt of computer files and manuscripts. The success of this workshop and of the early phases of the proceedings was in large measure assured by Diane's efficient dedication to these tasks and is gratefully acknowledged. Special thanks should also be passed on to Betty Alce, who took on the at times frustrating task of typing and formatting the manuscripts for publication. Her competence and patience are much appreciated by all.

Context

The Present State of GIS and Future Trends

Steven Reader[1]

Introduction

The term geographical information systems (GIS) has come to mean, variously, an industry, a product, a technology, and a science. As such, the term invokes different perceptions dependent on whether the viewpoint is that of the software developer, the system marketer, the data provider, the application specialist, or the academic researcher, among others. A newcomer to the field is likely to be bewildered by the multiple uses of the same term or, seen alternatively, different definitions of the same term.

In this paper, an effort is made to demystify usage of this term by drawing a broad distinction between GIS as an industry/ product/technology, and GIS as a science. In the former case, GIS is viewed very much as a technological tool which helps the analyst to use his/her knowledge and insight to study substantive issues. In the latter case, GIS is viewed as the science of geographical or spatial information which possesses its own set of research questions (see Rhind et al. 1991). The current state and future trends of each of these views of GIS is presented and the relationship between the two views is discussed.

GIS as Industry/Product/Technology

As an industry, GIS is commonly perceived as a specialized niche of "information systems." It is seen as bounded, in a porous sense, by computer aided design (CAD), remote sensing, and relational database management systems (RDBMS). Not unlike these and other information systems, the commercial world of GIS is viewed as an ill-defined combination of hardware, software, data, and consulting. Companies that operate within the industry tend to supply products and/or support services in one or more of these areas. Many of the companies in the hardware, data, and consultant sectors are major players in the much wider information systems industry, although there are specialized niches within these sectors which deal primarily with the spatial data handling industries (e.g., digitizing and large-format scanning hardware). However, the core of the industry, that which represents the purest primary GIS business (Dangermond 1991), is the multitude of GIS software vendors. Even here, however, many companies now

[1]Department of Geography, McMaster University, Hamilton, Ontario, Canada.

engage in the provision of hardware, data, consulting, and support services (e.g., training). As an industry, then, GIS is best viewed as a loose consortium of interests that has come into existence almost entirely within the last decade, and on the back of tremendous advances in computer processing and hardware technology. The industry growth rates for this period have probably been of the order of 25–35% per annum (Maguire 1991).

As a product, GIS is more readily defined as computer software, although so-called GIS "solutions", as sold by system vendors and consultants, may frequently involve hardware, database design/capture, and training. As a software product, the term GIS has often been loosely interpreted or, less graciously, manipulated, so that cartographic and mapping packages are commonly marketed as GIS. These packages generally lack both the analytical functionality and the well-developed links to RDBMS of GIS and, if vector-based, rely on simple data structures which ignore topology (Maguire 1991).

The lack of a rigorous definition of what constitutes GIS software has also led to confusion with image processing and CAD. Generally speaking, image processing packages specialize in the manipulation and display of raster data. Although they share much basic functionality with raster data handling in GIS, such packages also have sophisticated forms of analysis which go beyond those present in GIS. Image processing packages, however, typically have very limited capabilities in vector data handling and poorly developed links to RDBMS. CAD packages, not unlike computer cartography software, generally lack topology and also have weak relationships with RDBMS, although they are strong in 3-D data handling since they allow, unlike GIS, multiple surface values for the same x–y planar location (Maguire 1991).

Finally, a recent trend in GIS software has been the production by GIS vendors of specialized software or software modules for use in specific application areas. This trend towards exploiting vertically integrated markets is partly a reflection of the maturing of GIS as an industry. Progressive application areas (such as forestry or municipal affairs), now use GIS across a broad range of inventory, analysis, and management functions and, consequently, across many different departments within the same organizations. However, this trend also reflects the fact that users are increasingly looking for tailor-made and streamlined solutions to their work which sidestep the need to learn a full-blown GIS. As smaller GIS vendors have responded to this demand in certain sectors, so competitive forces have forced the major GIS vendors to respond likewise. In terms of the perception of what constitutes GIS, however, users of these specialized systems will undoubtedly have a restricted view and one which is interpreted solely in their own application context. One result of this trend has been a further proliferation of alternative names for GIS: cadastral information system, market analysis information system, soil information system, spatial decision support system, and so on.

As a technology, GIS transcends disciplinary boundaries and has found wide acceptability across a range of application areas, including land use management, traffic routing/assignment, political redistricting, resource management, and environmental modelling. The widespread use of GIS has much to do with the acceptance of the map as a means of communication, in addition to developments in graphical computing. These developments have enabled the map medium to be presented, manipulated, and analyzed in a new form and with unparalleled flexibility. GIS allows us to approach spatial data handling with much improved efficiency in implementing traditional methodologies and techniques.

Nowhere is this more evident than in the quintessentially GIS operation of map overlay. Although a relatively simple notion to comprehend and a formal analysis technique which dates back at least to McHarg (1969), this ability to link information together across numerous thematic layers for the same location and at great speed is an incredibly powerful tool. Whereas traditional RDBMS capabilities merely retrieve information records based on key fields, map overlay represents a major extension in that new data records, representing derived spatial features, may be created and populated with information drawn from different themes. Other traditional methodologies which benefit from the data structures employed in GIS are buffer operations (particularly in raster processing) and network analysis (vector processing).

However, despite the obvious power of GIS to manipulate and integrate data and to perform certain kinds of spatial analysis, many agencies use GIS merely for inventory management (Rhind 1988; Dangermond 1991). It seems that the relative novelty and power of working in a graphical environment for the management of information, and the satisfaction of being able to produce maps of that information at will, has perhaps obscured the vision of users from the real potential of suitably developed GIS for in-depth spatial analysis.

In turn, the willingness of agencies to purchase GIS purely on the basis of this automation of current spatial data handling and spatial data inventory, has led to few incentives for commercial GIS software vendors to incorporate spatial (and temporal) analytic functionality into their systems. Given the present sophistication of statistical packages for analyzing non-spatial data, it is not surprising that the existing state of GIS analytical capability is often compared to the state of statistical packages in the early 1970s (Rhind et al. 1991).

Regardless of the level of analysis, GIS is typically seen as a technological tool which helps the analyst use his/her knowledge and insight to study substantive issues. This viewpoint places GIS firmly in the role of an enabling technology. Although there is little doubt that GIS fulfils this role to a large degree, it conveys the impression that GIS merely provides a toolbox for operationalizing some form of analysis focused around a substantive issue. The implication is that analysts needs to be all-knowing with regard to their substantive field and that issues surrounding the enabling techniques play a very secondary role. In this sense, GIS has a lot in common with statistics use: analysts all too frequently use inappropriate statistical methods or correct statistical methods inappropriately.

There also seems to be a perception that the mere viewing of information and data that GIS permits is sufficient, and that this simply stimulates the user's own substantive knowledge which then takes over. Although this role is undoubtedly useful, this view tends to stifle the development of techniques to help the user detect and interpret patterns more objectively.

It can be argued, then, that GIS should in fact be the science of geographical information, and should concern itself with fundamental research issues of using digital geographical data. In the same way that we accept the discipline of Statistics for its fundamental research, and recognize how development at the research level feeds into the statistical software technology that is used by countless analysts, so there should be little problem accepting a similar situation for GIS.

Indeed, were it not for the heavy computational demands of graphical geographical data which delayed the development of GIS software relative to that of statistical software, we may well have seen Geography become to GIS software what Statistics has become to the software industry it supports. Instead, and somewhat paradoxically, geography went through an essentially a spatial quantitative revolution, often at the expense of those parts of the discipline, notably cartography, which were interested in issues of handling digital spatial data.

In the academic setting of today, it is difficult to see the diverse discipline of Geography focus itself purely as the science of geographical information. Meanwhile, the rapid and ubiquitous growth of GIS technology, partly based on the appeal of a highly graphical computing environment and the popularity of the map as media, makes potential misuse and abuse a significantly large problem. The need for a science of geographical information, then, is very real (Goodchild 1990), and perhaps merits a discipline in its own right.

GIS as Science

As a science, there is no shortage of basic research questions which underlie GIS and transcend the particulars of the technology and its applications (see Rhind et al. 1991). These include such questions as data capture, data accuracy, data volume, and generalization, all of which are crucial to any form of digital representation.

There also exists a need to develop appropriate data structures and data models for the handling of 3-D and temporal data within GIS, and legitimate and substantial research questions surrounding the use of expert systems and artificial intelligence. To these issues, we could add the need to develop database management systems and query languages which are geared towards spatial data, and the particular concern surrounding the issue of error propagation and error management within GIS.

Finally, there is the extremely important question of the relationship between spatial analysis and GIS. We need to address the problems of integrating existing spatial analytical methods into GIS, and of developing new methods of spatial analysis which specifically take advantage of the data structures within GIS. The importance of these issues becomes apparent if we consider the impact that scientific ideas such as the TIN data structure or the quadtree have had on the technological development of GIS.

A review of each of these areas is clearly beyond the limits of this paper, but the reader is referred to the excellent two-volume publication by Maguire et al. (1991) for papers which cover these topics. For the purposes of this paper, the issue of spatial analysis and GIS will be examined more closely. The justification for this focus lies in the fact that, ultimately, the real value of GIS will be in solving complex problems using sound and rigorous methods which are firmly rooted in spatial statistical theory. It is significant that the issue of spatial analysis and its relationship to GIS is a key research issue for both the U.S. National Centre for Geographical Information and Analysis (NCGIA 1989) and the U.K. Regional Research Laboratory initiative (Masser 1988).

Fotheringham and Rogerson (1994) have recently produced an edited volume which focuses on spatial analysis and GIS. In a review chapter, Bailey (1994) makes a useful distinction between spatial summarization of data, and spatial analysis of data. The former is meant to include functions for the selective retrieval of spatial information and the computation, tabulation or mapping of statistical summaries of that information. In Bailey's terms, this category of functionality would include both spatial query and many other techniques, such as boolean operations, map overlay, and buffer generation, which many users commonly perceive as analysis functions.

Meanwhile, the term spatial analysis is reserved for methods which either investigate patterns in spatial data and seek to find relationships between such patterns and the spatial (and perhaps temporal) variation of other attributes, or for methods of spatial or spatio–temporal modelling. The second of these would include network analysis, location-allocation models, site selection, and transportation models, all of which are considered by Bailey to be quite well developed within many GIS. The former type of spatial analysis, however, which Bailey refers to as statistical spatial analysis or simply spatial statistics, is currently poorly represented in the technology of GIS. This type of analysis would include such areas as nearest neighbour methods and K-functions, Kernel and Bayesian smoothing methods, spatial autocorrelation, spatial econometric modelling, and spatial general linear models. The level of integration of these types of methods with GIS has barely gone beyond the use of GIS to select input data and display model results. At the software level, this same integration generally involves only loose coupling between GIS and spatial statistical packages in the form of ASCII data transfer or specially programmed interfaces.

Bailey (1994) is somewhat pessimistic about the prospects of fully integrating statistical spatial analysis into GIS. He advocates a form of loose coupling, based on open-systems computing environments wherein the GIS package and the statistical analysis package would be accessed simultaneously but independently on the same GUI. The key problem would then be the data transfer mechanism between the packages. Bailey's pessimism centres on the number of theoretical problems which remain with spatial analysis methods, making them, in his opinion, difficult to implement in a sufficiently general and robust form for widespread consumption in commercial GIS. This underlines the necessity of developing a science of geographical information. Efforts would be better directed to addressing these fundamental problems, rather than accepting them as given.

GIS: The Future

Two views of GIS have been portrayed. For those who use the technology for spatial inventory management and mapping, GIS are perceived as an industry built upon a well-defined technology. This perception is also common to those who are attempting to familiarize themselves with the technology. Simply put, GIS are societal tools for use in many diverse application areas.

A second view perceives GIS as the science of geographical information, with the technology well defined. These two views should not be seen as competing alternatives, but rather as coexisting in the same way that the discipline of Statistics coexists with the statistical software industry. When discussing the future of GIS, we should include both views, and draw attention to the fact that the future of the technology depends partly upon the science, and the future of the science depends partly upon the technology.

As a technology, the future of GIS hardware seems easy to predict. There appears to be every indication that the rate of growth in computer processing power is likely to continue for the foreseeable future. Indeed, with parallel processing technology, this rate of growth may increase substantially, particularly for applications which can take advantage of it. This is definitely the case in GIS: the same operation may be carried out across a whole set of spatial objects, such as pixels. Tremendous ramifications for GIS will also result from hardware advances in global positioning systems (GPS).

These systems are already more accurate than the typical base mapping scales of most countries, and will redefine many of the methods of data capture and accepted levels of accuracy. Finally, GIS will be profoundly affected by the inevitable trend towards multimedia and networking. The ability to display text, maps, data, photographs, video, and sound for locations from a myriad of networked sources will give new definitions to what we typically think of as a spatial database.

Beyond hardware, there is also little doubt that GIS in the future will show increasing specialization. GIS application areas have very different requirements in terms of level of functionality, speed of response, and quantity of data handled. For instance, the GIS requirements of an emergency dispatch system, for which time is of the essence, have little in common with the management of cadastral land parcels, for which data volume may be the main concern. As specialized products emerge, the possibility exists that GIS as a distinct and discernible field will disappear.

Countering this trend, however, will be a unifying concern with standards, including the problems of data definition, function definition, data accuracy, and data exchange. As principal suppliers of digital data for GIS, public agencies have a major role to play in these areas; their capacity and willingness to do so is a critical question for the future. The one caveat in this regard, however, may be the propensity for governments to use the excuse of data standards to justify treating spatial data as a commodity, thereby demanding heavy payment and restrictive usage agreements.

The GIS industry may expand significantly, particularly at the low-end, with what would essentially become graphical interfaces to spatial databases. Good examples of this trend include the product ArcView from ESRI, the market leader in GIS software, and the addition of spatial data handling capabilities to such packages as Lotus 1-2-3 and SAS. GIS may actually become as commonplace in the desktop computing world as word-processing, spreadsheets or database packages. Moreover, they would feature the same embedding and linking technologies (e.g., OLE) that are currently integrating these different software platforms seamlessly.

Again, but for different reasons, the availability of GIS in such a format may actually dilute the identity of the field as users simply accept such capabilities as commonplace and come to expect them as an inherent part of any desktop software suite. The enthusiasm with which ESRI is attempting to introduce ArcView to the library and K-12 school sectors in North America represents an important portent.

As a science, a number of possible scenarios also exist for GIS. GIS must increasingly become a discipline in its own right, although not necessarily under that banner. The issues for spatial data analysis have been widely documented (see Openshaw 1991a; Goodchild et al. 1992), and there is a legitimate case for creating a separate scientific discipline. Already, GIS has many of the features of a separate discipline in journals, conferences, research institutes, and degree programs, albeit embryonic and often under the umbrella of a parent discipline, such as Geography.

The extent to which this could develop further is open to question, particularly given the range of parent disciplines from which academics interested in this new science would be drawn, including computer science, ecology, geography, geology, economics, and environmental science. Also, in the present academic climate of encouraging interdisciplinary work in areas which cut across traditional disciplinary boundaries, such as health and the environment, it is unlikely that such a new discipline would be recognized.

However, if GIS as science is to continue to garner academic recognition and funding, it is perhaps vital that the science be able to demonstrate the importance and usefulness of its research to GIS applications (see Goodchild et al. 1992) so that the wider GIS industry becomes an advocate and potential financial partner. This will involve a critical assessment by researchers of spatial analysis methods and their relevance to the powerful computing and GIS environment of the 1990s. In this regard, it should be noted that much of the research in spatial analysis methods dates back to the 1960s and early 1970s, followed by a relative dearth of research until the GIS-inspired revival of the late 1980s and 1990s. As Openshaw (1991b) argues, many methods from the early research were developed within a completely different computing environment. It may be that new methods, which capitalize on the raw computing power available today and into the future, should now be the focus of attention.

Another scenario for GIS as a science, could be that traditional disciplines will struggle with the issues of spatial data handling and analysis somewhat independently. In this scenario, GIS as science may become little more than a very loose and informal consortium of academic interests, much like the field of remote sensing. The dangers of research duplication and dilution would be very real, and a lack of critical mass in the form of academic institutions to spearhead the adoption of methods into GIS would result. This scenario, then, may be a recipe for the field of GIS being led too rapidly by technology and with science not being given the chance to catch up, a familiar theme to the field of remote sensing.

Conclusion

The field of GIS may be at a critical juncture in its development. On the one hand, an extremely wide awareness exists of the technology, and the numbers and range of adopters are rapidly growing. Continued improvements in hardware and the increasingly competitive nature of the GIS software industry seem destined to fuel this growth far into the future. There is also a growing contingent of mature GIS application areas which possess well-developed spatial databases and a body of experienced users.

On the other hand, a belief exists that the typical use of GIS has not progressed far beyond the use of mapping, query, and spatial data inventory management, and that the potential analytic power of the technology to help solve complex societal and environmental problems has yet to be realized. For this to happen, there is a need for fundamental research into the science of geographic information, a need for more widespread and enhanced education in this science, and a willingness on the part of the GIS industry to nurture this science and be ready to adopt and promote the analytical techniques it produces.

References

Bailey, T. 1994. A review of statistical spatial analysis in geographical informtion systems, in spatial analysis and GIS (Fotheringham, S.F. and Rogerson, P., ed). Taylor and Francis, London, UK. pp.13–44.

Dangermond, J. 1991. The commercial setting of GIS, in geographical information systems. Volume 1: Principles (Maguire, D.J. et al.). Longman. London, UK. pp. 55–65.

Fotheringham, S.; Rogerson, P. 1994. eds. Spatial analysis and GIS. Taylor and Francis, London, UK.

Goodchild. M.F. 1990. Spatial information science. Proceedings of the 4th International Spatial Data Handling Symposium. International Geographical Union, OH, USA. pp. 3–14.

Goodchild, M.F.; Haining, R.; Wise, S.; 12 others. 1992. Integrating GIS and spatial data analysis: Problems and possibilities. *International Journal of Geographical Information System*, 6, 407–423.

Maguire, D.J. 1991. An overview and definition of GIS, in geographical information systems. Volume 1: Principles (Maguire, D.J. et al.). Longman, London, UK. pp. 9–20.

Maguire, D.J.; Goodchild, M.F.; Rhind, D.W. 1991. Geographical information systems. Longman, London, UK.

Masser, I. 1988. The regional research laboratory initiative. *Geographical Information Systems International Journal* of G 2, 11–22.

McHarg I.L. 1969. Design with nature. Doubleday, New York, NY, USA.

NCGIA. 1989. The research plan of the National Centre for geographic information and analysis. *International Journal of Geographical Information Systems*, 3, 117–136.

Openshaw, S. 1991a. Developing appropriate spatial analysis methods for GIS, in geographical information systems. Longman, London, UK. pp. 389–402.

Openshaw, S. 1991b. A spatial analysis research agenda, in handling geographical information (Masser, I. and Blakemore, M., ed.) Longman, New York, NY, USA. pp. 18–37.

Rhind, D.W. 1988. A GIS research agenda. *International Journal of Geographical Information Systems*, 2, 23–28.

Rhind, D.W.; Goodchild, M.F.; Maguire, D.J. 1991. Epilogue in geographical information systems. Volume 2. Applications (Maguire, D.J., et al.). Longman, London, UK. pp. 313–327.

Geographical Information Systems (GIS) from a Health Perspective

Luc Loslier[1]

Introduction

A GIS can be defined as a computer-assisted information management system of geo-referenced data. This system integrates the acquisition, storage, analysis, and display of geographic data. The application field and objectives of a GIS can be varied, and concern a great number of questions linking social and physical problems (transport and agricultural planning, environment and natural resources management, location/allocation decisions, facilities and service planning (education, police, water, and sanitation), and marketing).

Generally speaking, the objectives of a GIS are the management (acquisition, storage, maintenance), analysis (statistical, spatial modelling), and display (graphics, mapping) of geographic data. Even if a few general concepts are presented, the GIS discussed here will be seen from a health perspective. Thus, GIS will be considered as a tool to assist in health research, in health education, and in the planning, monitoring, and evaluation of health programs.

Geographic Information Systems and Health

A GIS can be a useful tool for health researchers and planners because, as expressed by Scholten and Lepper (1991),

> Health and ill-health are affected by a variety of life-style and environmental factors, including where people live. Characteristics of these locations (including socio-demographic and environmental exposure) offer a valuable source for epidemiological research studies on health and the environment. Health and ill-health always have a spatial dimension therefore. More than a century ago, epidemiologists and other medical scientists began to explore the potential of maps for understanding the spatial dynamics of disease.

A study carried out by John Snow is often cited to show that the importance of spatial dynamics in the understanding of disease, and the use of maps to describe and analyze it, is not so recent. Dr Snow made the hypothesis that cholera might be spread by infected water supplies more than a century ago,

---

[1]Université du Quebéc à Montréal, Montréal, Québec, Canada.

using maps to demonstrate in a striking fashion the spatial correlation between cholera deaths and contaminated water supplies in the area of Soho in 1854.

Scholten and Lepper use the example of AIDS, stressing the importance of the spatial distribution of the disease, which they say has been too often overlooked. They cite Kabel (1990):

> modelling the spatial distribution of AIDS can contribute to both educational intervention and the planning of health care delivery systems. Mapping can play an important role in both areas as it is an excellent means of communication. In order to be of use to resource planners, predictions of AIDS should include a spatial component.

The Database

The database is central to the GIS, and contains two main types of data. There are in fact two databases (more or less closely integrated, depending on the system): there is a spatial database, containing locational data and describing the geography of earth surface features (shape, position), and there is an attribute database, containing certain characteristics of the spatial features.

The Spatial Database

The information contained in the spatial database is held in the form of digital coordinates, which describe the spatial features. These can be points (for example, hospitals), lines (for example, roads) or polygons (for example, administrative districts). Normally, the different sets of data will be held as separate layers, which can be combined in a number of different ways for analysis or map production.

The Attribute Database

The attribute database is of a more conventional type; it contains data describing characteristics or qualities of the spatial features: land use, type of soil, distance from the regional centre, or, using the same example as in the preceding paragraph, number of beds in the hospital, type of road, population of the administrative districts. Thus, we could have health districts (polygons) and health care centres (points) in the spatial database, and characteristics of these features in the attribute data base, for instance persons having access to clean water, number of births, number of one year-old children fully immunized, number of health personnel, and so on.

Data Input Systems

The Digitizing System

One important source of locational data is existing paper maps, for example, road maps or administrative boundaries maps. The digitizing system that is part of most GIS allows one to take these paper maps and convert them into digital form (this is not necessarily done by the GIS end user; it is often produced by another party).

The Image Processing System

Another source of data for a GIS is remotely sensed imagery, such as LANDSAT or SPOT satellite imagery. A complete GIS offers tools to convert raw remotely sensed imagery into maps. Thus, an enormous quantity of environmental data pertinent to health can be integrated into a health-oriented GIS (like digitizing, this task is not necessarily done by the GIS end user).

Other Data Input Methods

GIS have interfaces that permit the importation of data from numerous database or worksheet programs. Image collection devices, such as scanners, cameras, or tape players, can also transfer images from paper or photographic materials (maps or aerial photographs, for example) into the GIS database.

The Cartographic Display System

The cartographic display system is the map producing tool. It allows the user to extract necessary elements from the database, such as spatial features and attributes, and to rapidly produce map outputs on the screen or other devices, such as high speed electro-static plotters or simpler pen-plotters, laser printers, or graphic files in popular formats.

The Database Management System

The database management system is used for the creation, maintenance, and accessing of the GIS database. The system incorporates the traditional relational database management system (RDBMS) functions, as well as a variety of other utilities to manage the geographic data. The traditional database management system makes it possible to pose complex queries, and to produce statistical summaries and tabular reports of attribute data. It also provides the user with the ability to make map analyses, often combining elements from many layers. For instance, a user might ask the system to produce a map of all districts where a health centre exists and where the proportion of 0–1 year-old children who received required vaccination is less than 50%. For a problem like this, the map analysis does not have any actual spatial component, and a traditional DBMS can function quite well. "The final product (a map) is certainly spatial, but the analyses itself have no spatial qualities whatsoever" (Eastman 1992).

A powerful, relational DBMS is a requisite part of a GIS, to handle large quantities of information. It can provide very useful results, but a GIS must have another set of tools to give it the ability to analyze data based on their spatial characteristics. This set of tools corresponds to the geographic analysis system.

The Geographic Analysis System

A variety of analytical tools are available within GIS, extending the capabilities of traditional DBMS to include the ability to analyze data based on their spatial characteristics.

The Overlay Process

Eastman (1992) gives an example of the ability of GIS to analyze data based on their spatial characteristics:

> Perhaps the simplest example of this is to consider what happens when we are concerned with the joint occurrence of features with different geographies. For example, find all areas of residential land on bedrock types associated with radon gas. This is a problem that a traditional DBMS simply cannot solve — for the reason that bedrock types and landuse divisions simply do not share the same geography. Traditional data base query is fine so long as we are talking about attributes belonging to the same individuals. But when the entities are different it simply cannot cope. For this we need a GIS. In fact, it is this ability to compare different entities based on their common geographic occurrence that is the hallmark of GIS — a process called "overlay" since it is identical in character to overlaying transparent maps of the two entity groups on top of one another.

The example given in the preceding paragraph can be developed here: Let us say we want again a map showing the districts where there is a health centre and where less than 50% of 0–1 year-old children have received necessary vaccination. We want also to have on that map the hydrographic system (lakes, ponds, and rivers) of the area and the location of clean water sources and sanitation utilities. We need a GIS, because the immunization data, the water and sanitation data, and the hydrographic system data have different geographies. The analytical tools available within GIS are necessary to make possible the integration of data having different geographies.

It should be noted that the geographic analysis system can contribute to the extension of the database; for example, by combining the areas where the immunization rate is low and the access to clean water is difficult, the analyst defines zones and populations at greater risk. In this way, new knowledge of relationships between features is added in the database.

Buffer Zones Creation

The overlay process is among the most fundamental aspects of a GIS, but other processes are important and can be very useful in health research and planning.

> Of particular benefit to the investigation of illness at or near pollution and other hazardous sites is the ability to create buffer zones around the lines or points which represent those locations. The user can specify the size of the buffer and then intersect or merge this information with disease incidence data to determine how many counts of the illness fall within the buffer (Twigg 1990).

The association between proximity to nuclear power stations and the prevalence of childhood leukemia in northern England (Openshaw et al. 1987) has been investigated in this way, and one can easily imagine similar applications with other diseases and other environmental causes or risk factors of disease.

Buffer zones analysis can have useful applications in health services analysis and planning; for example, it gives a quick and easy answer to the question: "How many persons live within a 10 km radius from this health care centre? Within a 10–15 km radius?" The generation of a distance/proximity surface (taking into account distance and "friction" of space — resulting in a cost (in money or time...) of transportation) and allocation modelling (assignment of every point of an area to the nearest of a set of designated features (for example health centres)) are other geographic analysis tools that can be useful in health research and planning, where a nonspatial method could give a partial or even false answer.

For example, a GIS was used to study the difference in population per bed ratios between blacks and whites, and the implications of open access to hospital services formerly reserved for whites in Natal, South Africa. While the usual administrative boundary-based beds per capita ratios suggested that hospital bed resources in the province of Natal/Kwazulu were racially unequal but, nevertheless, as expressed by Zwarenstein et al. (1991):

> adequate (264 people per general and referred bed for the whole population, 195 for whites and 275 for Blacks), the GIS analysis reveals widespread inadequacy, worse for blacks. Of the estimated hospital catchment areas half have more than 275 black people per general and referral bed, and half of these have more than 550 black people per bed.

One-third of the catchment areas estimated for whites have ratios above 275 people per bed and one half of these are also above 550 persons per bed. The GIS analysis shows that open access to beds previously reserved for whites will make no difference to rural blacks, and almost none to urban blacks, because there were relatively few such beds, and they were concentrated in the cities. For the same reasons, the opening of private hospital beds would not alleviate the apparent bed shortages in priority areas.

Conclusion

As health is largely determined by environmental factors (including the sociocultural and physical environment, which vary greatly in space), it always has an important environmental and spatial dimension. The spatial modelling capacities offered by GIS can help one understand the spatial variation in the incidence of disease, and its covariation with environmental factors and the health care system. GIS in health-related activities can play a role at three levels:

GIS and Health Research

By helping to understand the distribution and diffusion of disease and its relationship to environmental factors (climate, water quality, sanitation, land use, agricultural, and other economic activities, rural–urban milieu, immunization rate, and so on), it is of value to etiology, epidemiology, and medical science in general.

GIS and Health Education

As mapping is an excellent means of communication, GIS can be used, as Kabel (1990) suggests, to help prepare educational material. In an article on participatory evaluation, Fuerstein (1987) describes different methods for monitoring and evaluating community health projects, including mapping.

> Small or large maps may be drawn or painted by groups or individuals to represent the context in which they are living (...) These maps, showing location of houses by number and type, public and private buildings, water sources, sanitation, bridges, roads, social centres, neighbourhood boundaries, health centres, etc. give participants a wider view of where they are living. Maps can help discussion, analysis, decision-making, management and evaluation.

Fuerstein suggests that these maps be posted in a public place and updated as changes occur, providing a permanent record. GIS thus produce material which is both useful and conducive to public participation in community health projects. GIS can contribute to community development in general, by helping people understand their environment. Effects in the health domain are obvious. From this perspective, indicators developed with the people, such as the 32 indicators found in the Basic Minimum Needs (BMN) database of Thailand (Nondasuta and Chical 1988), or those measuring the 30 priority problems identified in the Recherche nationale essentielle en santé program in Bénin (Badou 1994), deserve special attention (these indicators reflect the health level as well as social, economic, and environmental key determinants of health).

GIS and Health Planning

It is evident that many questions concerning the provision of health care are related to space. People are distributed in space and they are not evenly distributed. Health problems vary in space and so do the needs of the people.

Where should health care centres be situated and what services should they offer to answer efficiently to the needs of populations varying in numbers, densities, and health problems? These are problems that GIS can help resolve with their spatial analysis tools.

Maps produced by a GIS can also be used by health officials as a monitoring and evaluation tool, showing the spatial distribution and differential evolution of diseases. Monitoring and evaluation are essential parts of health programs, as well as other programs related to development. As the WHO/UNICEF Joint Monitoring Program points out (1993),

> Monitoring is defined as the periodic oversight of the implementation of an activity which seeks to establish the extent to which input deliveries, work schedules, other required actions and targeted outputs are proceeding according to plan, so that timely action can be taken to correct the deficiencies detected.
>
> Closely linked to monitoring is evaluation. Evaluation is a process by which program inputs, activities, and results are analyzed and judged explicitly against stated norms. These two terms are usually used in tandem as an integral part of every program.
>
> (...)Monitoring is an essential element. By giving the managers, planners and policy-makers access to information on coverage, functioning and utilization of the Water and Sanitation facilities, operation and maintenance, funding, water quality and others, monitoring as a tool guides them in making important decisions. Similarly, worthwhile evaluation of water and sanitation, as a result of effective monitoring, is necessary in ensuring rational utilization of investments allocated for the sector.

It is worth noting that the second version of the WASAMS (Water and Sanitation Monitoring System) software contains a feature developed to facilitate the link with GIS. The WHO/UNICEF Joint Monitoring Program's comments on the water and sanitation program could be said, with the same words, about health programs. Monitoring and evaluation are essential parts of health programs and GIS, by showing the spatial distribution of diseases in space and time, facilitate the monitoring and appraisal of the effectiveness of health programs. GIS are a relatively recent and complex technology, which explains why they have not been used to their full potential, especially in the health domain where they are extremely promising. We are now to a point where their possibilities are more clearly seen. Hardware and software development has produced systems with functions and interfaces which make them much easier to use. This is very good news, as GIS can certainly be a tool of prime importance to health research and education, and in the planning, monitoring, and evaluation of health programs.

References

Badou, J.A. 1994. Les populations béninoises mobilisées. Le CRDI explore, 22(1).

Eastman, J.R. 1992. IDRISI User's Guide, Version 4.0 rev.1. Clark University Graduate School of Geography, Worcester, MA, USA.

Fuerstein, M.T. 1987. Partners in evaluation. leadership. ASEAN Training Centre for Primary Health Care Development, 1(1), 34–35.

Kabel, R. 1990. Predicting the next map with spatial adaptive filtering. In Proceedings of the fourth international symposium in medical geography. Norwich, 16–19 July Norwich, University of East Anglia, UK.

Nondasuta, A.; Chical, R. 1988. The basic minimum needs guiding principles the foundation for quality of life. Ministry of Public Health. Bangkok, TH..

Openshaw, S.; Charlton, M.; Wymer, C.; Craft, A. 1987. Building a Mark 1 geographical analysis machine for the automated analysis of point pattern cancer and other spatial data. Economic and Social Research Council Northern Regional Research Laboratory Research Report No. 12. University of Newcastle upon Tyne, UK.

Scholten, H.J.; de Lepper, M.J.C. 1991. The benefits of the application of geographical information systems in public and environmental health. *WHO Statistical Quarterly,* 44(3).

Twigg, Liz. 1990. Health base geographical information systems: Their potential examined in the light of existing data sources. *Social Science and Medicine,* 30(1).

WHO/UNICEF Joint Monitoring Program. 1993. Water and sanitation monitoring system, guide for managers, WHO, Geneva, Switzerland.

Zwarenstein, M.; Krige, D.; Wolff, B. 1991. The use of a geographical information system for hospital catchment area research in Natal/KwaZulu. *South African Medical Journal,* 80(10).

Spatial and Temporal Analysis of Epidemiological Data

Flavio Fonseca Nobre[1] and Marilia Sá Carvalho[2]

Introduction

Public health practice needs timely information on the course of disease and other health events to implement appropriate actions. Most epidemiological data have a location and time reference. Knowledge of the new information offered by spatial and temporal analysis will increase the potential for public health action. Geographical information systems (GIS) are an innovative technology ideal for generating this type of information.

Spatial analysis and the use of geographical information systems for health have been reviewed by several authors (Mayer 1983; Gesler 1986; Twigg 1990; Marshall 1991; Scholden and de Lepper 1991; Walter 1993). Of major interest has been detecting clusters of rare diseases, such as leukaemia near nuclear installations, methods for mapping and estimating patterns of disease, and health care location/allocation problems. Temporal analysis has focused primarily on the detection of cluster and abnormal case occurrence of notifiable diseases (Helfenstein 1986; Zeng et al. 1988; Stroup et al. 1989; Watier and Richardson 1991; Nobre and Stroup 1994). Work on spatial–temporal analysis has been more limited. Emphasis has been on the presentation of time series maps (Sanson et al. 1991; Carrat and Valleron 1992), the use of contour plots to analyze the time of the introduction of disease into households (Splaine et al. 1974; Angulo et al. 1979), and the use of specified contour levels extracted from the spatial distribution of a disease for different time frames (Sayers et al. 1977).

This paper is a survey of the concepts and methodologies of spatial and temporal analysis that could be beneficial for health-related studies, particularly to public health professionals and policymakers at the state, national, and local levels. Emphasis will be on integrative tools that can help in the construction of a geographical information system for epidemiology (GIS–EPI).

[1]Programa de Engenharia Biomédica, Universidade Federal do Rio de Janeiro, (COPPE/UFRJ), Rio de Janeiro, Brazil.

[2]Escola Nacional de Saúde Pública, FIOCRUZ, Brazil.

Spatial Analytic Techniques

Spatial variation in health related data is well known, and its study is a fundamental aspect of epidemiology. Representation and identification of spatial patterns play an important role in the formulation of public health policies. The spatial analytic techniques reviewed here are limited to those involving a graphical, exploratory analysis of data.

Point Patterns

As the name implies, point patterns, also known as dot maps, attempt to display the distribution of health events as data locations. The ability to overlay data locations with other relevant spatial information, such as a city map or the distribution of health facilities, is a general tool of considerable power. This is the simplest method of spatial analysis. It is useful for delimiting areas of case occurrences, identification of contaminated environmental sources, visual inspection of spatial clusters, and analyzing health care resources distribution.

A classical example of point pattern analysis in epidemiology, is the identification of the source of cholera spread in London by John Snow. In 1854, using dots representing cholera deaths in the Soho area of London, Snow identified the source of contamination as a water pump. It is not surprising that until now, most texts of health surveillance recommended the use of pins to locate cases of notifiable diseases on a map. Several statistical techniques have been suggested to study point patterns (Gesler 1986). These methods, however, are not yet integrated into a seamless software allowing easy access and use by health professionals at different levels.

An attractive alternative to point patterns is the use of dynamic graphics, which could be more easily implemented within a GIS software (Haslett 1991). Using this approach, the dot map can be associated with a histogram of case occurrences. Selecting the upper tail of the histogram automatically highlights the corresponding cases on the map, allowing the eventual characterization of regions with a high incidence of a disease. Alternatively, cases selected in one area of the map can be classified in the histogram. The temporal case occurrence for each data location could also be displayed.

Line Patterns

Vectors or lines are graphical resources which aid in the analysis of disease diffusion and patient-to-health care facilities flow. In their simplest form, lines indicate the presence of flow or contagion between two subregions that may or may not be contiguous. Arrows with widths proportional to the volume of flow between areas are important tools to evaluate the health care needs of different locations. Francis and Schneider (1984) have designed an interactive graphic computer program, called FLOWMAP, producing a variety of maps of origin–destination data. These maps have been used to assess referral patterns of cancer patients in the northwestern part of the state of Washington, USA.

Use of line pattern analysis is quite common in epidemiology to describe the diffusion of several epidemics, such as the international spread of AIDS. It seems that the presence of a module such as FLOWMAP in a GIS package would enhance the applicability of line pattern analysis in public health.

Area Patterns

The first stage of data analysis is to describe the available data sets through tables or one-dimensional graphics, such as the histogram. For spatial analysis, the obvious option is to present data on maps, with the variable of interest divided into classes or categories, and plotted using colours or hachures within each geographical unit, know as a choropleth map. A literature search on spatial analysis revealed that among 76 papers, 54% used the choropleth map as an analytical tool.

The most common maps use pre-specified classes of health events, or the mean and standard deviation of their distribution. The maps are usually represented using administrative boundaries such as counties, municipalities, health districts, and so on, where data is at which is usually collected. Major variables used for area pattern analyses are incidence rates, mortality rates, and standardized mortality ratio (SMR). The latter is most common in health atlases (Walter et al. 1991). At times, area pattern analysis uses statistical significance rather than raw data. Area pattern analysis also uses empirical Bayes estimates of the relative risk (Marshall 1991).

When two or more health-related variables are available in each areal unit, multivariate analysis, can synthesize the information. Recently, measures of coherency between two variables were used as the variable of interest to explain the spatial pattern of disease occurrence (Cliff et al. 1992).

Several criticisms have been raised against the direct use of SMR and p-values or statistical significance. Major criticism of SMR-based approach is based on the influence of population size. Unreliable estimates from areas with low populations mask the true spatial pattern, presenting extreme relative risk estimates which dominate the map. Use of p-values for the significance of the variable under study is considered uninformative. When computing p-values, areas with a modest increase of risk are likely to display extreme values solely due to the high population. This problem has led researchers to favour the development of empirical Bayes estimates for area pattern analysis.

An interesting alternative, which may be less demanding, is the use of steam-and-leaf plots to classify data before area pattern analysis. This approach is more intuitive and easier to use by health professionals, and presents another method of incorporating dynamic graphics into a GIS for use by health professionals. Through the display of distribution data using steam-and-leaf plots, it is possible to select appropriate schemes of classification before the production of choropleth maps.

Another criticism of choropleth maps lies in their limitation in using administrative boundaries for studies of health events. Large areas may dominate smaller ones, where estimates are usually more reliable. To overcome this limitation, two main suggestions have been proposed: use of symbolic representation, and map transformation. The former use circles or frames proportional to the magnitude of the variable (Dunn 1987); the latter involves modification of the boundaries of the regions according to size of the population (Selvin et al. 1988). However, transformed maps sometimes are difficult to interpret, and may conceal clusters of health events for areas with larger populations when they are augmented.

Surface and Contour Patterns

Data of epidemiological or public health interest often occur as spatial information during each of several time epochs. The analytical techniques described previously require the pooling of information in administrative areas with well defined geographic boundaries (e.g., counties, municipalities, and health districts), and the presentation of the spatial process with maps constrained to them. These maps are often unable to capture health problems at the locality or subcounty level. As well, epidemiological variables do not necessarily recognize political boundaries. To overcome the limitation of administrative regions for mapping, surface and contour pattern analysis presents an alternative by representing the distribution of the health event. The advantage of this spatial analytical technique is that the variable under study is treated as a continuous process throughout the region.

Surface and contour analysis assumes that a health event is a continuous process observed at a set of geographic points, known as sampling points. Using the x and y coordinates of these sampling points, with an associate z value corresponding to the health event, the estimated spatial relative risk is depicted as a three-dimensional map or surface. The contour map, known as an isoline or isopleth, is the projection of the surface in a plane, and corresponds to constant z values of the defined surface.

Although these techniques may overcome the limitations of political boundaries and help in the representation of spatial processes collected as point data, they are not often used by health researchers. This may be due to the very fact that they lose the geopolitical information known to the researcher. One possibility, which is already available in some GIS packages, is the capacity of overlaying the geographical map of the region with one of these analytical techniques.

Temporal Analytic Techniques

Surveillance is usually understood as the continuous systematic collection and analysis of a series of quantitative measurements. The detection and interpretation of changes in the pattern of the constructed time series are very important. The surveillance of diseases, and of other health events, presents an important challenge to public health surveillance systems, since late detection of increases in case occurrence may result in missed opportunities for intervention. Temporal analytic techniques reported here centre on procedures for timely detection of the onset of a potential epidemic period. Other temporal analytic techniques, such as correlation analysis, will not be discussed.

Quality Control Charts

Industrial quality control has developed a series of methods for monitoring. Among them, three major methods appeared in the public health surveillance literature — the Shewhart test, the simple cumulative sum test, and V-mask. These methods are based on a comparison of incoming values from the time series with constant values, usually defined empirically from historical data. The advantages of these methods are that they can provide graphical information, and as such can be incorporated into an information system, helping public health professionals in the surveillance process.

The Shewhart test uses only the last observation, comparing it with a predefined target or expected value. An abnormal case occurrence is declared whenever the absolute difference between the observed and the expected value exceeds a specified threshold. The simple cumulative sum test uses the accumulated deviations between expected and observed values. Here, identification of abnormal events occurs when the absolute cumulative sum exceeds a fixed limiting value. Finally, the V-mask, which is also known as CUSUM method, is based on the construction of a graphic of the cumulated sums of the deviations between observed and expected value. For this method, a V-shaped mask moved over the resulting cumulative time series permits detection when an earlier observation crosses one leg of the mask.

A recent evaluation of the three methods suggests that in addition to the constant values used as an alarm threshold, other parameters of the methods should be provided to facilitate an understanding of the alarms (Frisén 1992). Also, application of quality control charts can be made on suitably transformed data.

Statistical Monitoring

A common measure used by epidemiologists to identify increases in case occurrence of diseases, is the ratio of case numbers at a particular time to past case occurrence using the mean or median. Based on this concept, a monitoring technique has been developed and is currently in use at CDC (Centres for Disease Control and Prevention, USA) (Stroup et al. 1989). Expected values for the current month are computed as the average of data from the corresponding, previous, and subsequent months for the last five years. The basic assumption of the method is that the obtained 15 values are independent random variables with the same distribution function.

The test statistic is the ratio of the observed value to the constructed average. Confidence intervals are computed and a comparison of actual and expected values is undertaken. A bar plot displays simultaneously the ratios of several notifiable diseases. Deviations from the unit of this ratio indicate a departure from past patterns, and values that exceed the 95% limits are shaded differently.

Although the method is quite simple to compute, it requires the availability of 5 years of past data. Also, problems due to correlate data values reduce the precision of the estimated confidence interval. This latter problem can be modified using some special techniques to estimate the variance (Kafadar and Stroup 1992).

The requirement of a 5-year period precludes the use of this technique in places where a surveillance system is starting, or for new health conditions. One advantage for public health officials, is the simplicity of the method and the straightforward form of presentation for final analysis.

Time Series Analysis

To account for the evolving nature of surveillance data, time series analysis is an alternative for monitoring case occurrence of health events. The common analytical framework uses time series models to forecast expected numbers of cases, followed by comparison with the actual observation. Detection of changes from historical patterns through forecast error uses the difference between the actual and estimated value at each point in time. In contrast to other monitoring schemes, time series methods use the correlation structure of the data at different time intervals in making estimates. Most attention has been focused on the use of the Box–Jenkins modeling strategy to construct autoregressive integrated moving average (ARIMA) models for specific health variables (Watier and Richardson 1991; Stroup et al. 1988; Helfenstein 1986). The modelling strategy analyzes a long series of values in a stationary mode. Since most health variables of interest are not stationary, the analysts have to resort to preliminary transformations, such as time series differencing or variance-stabilizing to achieve stationarity. After choosing the transformation, the steps of model identification, parameter estimation, and diagnostic checking are performed. Key tools for modelling are the autocorrelation function (ACF) and the partial autocorrelation function (PACF).

Monitoring is usually achieved by computing the forecast error and using one of the quality control charts described previously. This approach is quite complex, demanding a thorough statistical knowledge for its development. It also requires the availability of long historical data. Several health time series are not modeled; even after applying transformations stationarity is not attained.

Another time series approach for forecasting uses the exponential smoothing technique. Here, available data is approximated by a polynomial model and the coefficients are computed as each new observation become available. The computed forecast error from this modelling approach behaves as a derivative of the time series, and forms the basis for detecting abnormal pattern occurrence (Nobre and Stroup 1994). This method is less demanding than other time series approaches, requiring less historical data for the initial set up. This method has also been applied in scenarios with several periods with no case occurrence, which usually does not satisfy the assumptions of other modelling techniques. The major constraint of this analytical technique is the empirical process to estimate the parameters.

Time series analysis has already been shown to be quite useful in different contexts for monitoring tasks. Its implementation into an integrated system for use in public health will lead to a better assessment of its impact and utility. It will also open the opportunity for additional studies, such as the influence of climatic and other environmental time series on the occurrence of health events.

Temporal Cluster Analysis

Detection of temporal clusters, understood as a change in the frequency of disease occurrence, is important to stimulate research into the causes, and to encourage the development of preventive strategies. Detection of increases in the rate of occurrence of a disease uses either the time interval of successive events, or the number of events on specified time intervals.

One method for identifying clusters is the scan statistic test (Naus 1965). This method consists of counting the number of cases in each possible time interval of fixed length. The largest number of cases in any such intervals is tested under the null hypothesis that this value is likely to occur in a case of no epidemic. Application of this method involves the assumption of a constant population at risk and a constant detection rate of cases. A modification of the method has been suggested to avoid the restrictive assumptions involved in the scan statistic (Weinstock 1981). Studies of temporal clusters based on the time interval between events have also been described in the literature (Sitter et al. 1990; Gouvea and Nobre 1991). These methods assume that the random time intervals of successive cases form an independent and identically distributed sequence of exponential random variables.

Although these methods are useful for detecting temporal clusters, they are not easily accepted by public health officials, who usually have little knowledge of statistical methods. Thus, incorporation of this type of temporal analytical technique into an integrated information system as a GIS–EPI requires careful planning and evaluation.

Spatio–Temporal Analytic Techniques

Space–time interaction among health events or between health events and environmental variables is as an important component for epidemiological studies and public health surveillance. The bulk of the development in spatio–temporal patterns of health problems has been based on modelling and simulation because of the paucity of available data sets (Marshall 1991). Similarly, with time series analysis, the basis of spatio–temporal analytical techniques is the assumption that observed spatial patterns arise from an underlying process. Modelling this underlying spatial processes allows for the study of disease diffusion process, and the estimation of linear spatial transfer functions which best transform a map at time t into that at time t+1.

One example of a spatio–temporal analysis of the spread of contagious disease, is the use of the date of the onset of *variola minor* in the dwellings of a small town in Brazil (Splaine et al. 1974; Angulo et al. 1979). Different contour maps were produced using the date of introduction of the disease into the dwelling as the dependent variable. The independent variable were the x-and-y coordinates of the dwelling localization. This type of analysis can be conducted if data at a low spatial scale is available, which may be feasible with an adequately designed GIS–EPI system.

Study of the spread of a disease over large areas has been the subject of a more recent study that used kriging to estimate the underlying spatial process of an influenza-like epidemic in France (Carrat and Valleron 1992). A sequence of contour maps was produced, characterizing the spread of disease in the country. This technique is a simple description of the spatio–temporal evolution of the disease, but it may prove to be quite powerful for establishing control measures.

Another type of spatio–temporal analytic technique involves estimating the probability density of case occurrence of a health event. Use of this method has allowed the description of the spatio–temporal spread of rabies epizootic in central Europe (Sayers et al. 1977). In this work, a contour map of relevant probability levels depicted the wavefront spread of the disease, and line pattern analysis described the limited set of trajectories of the foci's movements across the region. Use of the estimated probability density also allowed the construction of a linear transfer function to estimate maps of the distribution of case occurrence from previous time epochs. This technique involves a strong knowledge of statistical analysis. While producing some interesting results, it requires additional evaluation before application by public health professionals.

Conclusions

This paper has outlined the concepts and functions of spatial and temporal analytical techniques and suggested their relevance for health care. Although these methods offer much potential, their use for health is crucially dependent upon the availability of suitable data. Also, it is fundamental that a GIS–EPI is developed to allow the data and the associated analytic results more accessibility and understandability to public health officials.

GIS may play an important role in the use and analysis of public health data. However, turning the promise into reality entails a multidisciplinary effort to explore the possibilities offered by spatial and temporal analytical techniques to improve our knowledge of public and environmental health. The issue of reliability and validity of the data is also of importance. Special attention should focus on ensuring that geographical information is available in digital form, and that different data sets can be linked accordingly.

Acknowledgments

The International Development Research Centre (IDRC) provides financial support for the Biomedical Engineering Program, PEB/COPPE/UFRJ through the research grant IDRC 92-0228. In addition, FFN and MSC received additional financial support from the Brazilian Research Council, (CNPq).

References

Angulo, J.J; Pedemeiras, C.A.A; Sakuma, M.E.; Takiguti, C.K.; Megale, P. 1979. Contour mapping of the temporal–spatial progression of a contagious disease. *Bull. Soc. Pathologie Exotique*, 374–385.

Carrat, F.; Valleron, A.J. 1992. Epidemiologic mapping using the kriging method: Application to an influenza-like illness epidemic in France. *Am. J. Epidemiol.*, 135, 1293–1300.

Cliff, A.D.; Haggett, P.; Stroup, D.F.; Cheney, E. 1992. The changing geographical coherence of measles morbidity in the United States. 1962–88. *Stat. Med.*, 11, 1409–1424.

Dunn, R. 1987. Statistical mapping. *The Am. Statistician*, 41, 153–156.

Frisén, M. 1992. Evaluation of methods for statistical surveillance. *Stat. Med.*, 11, 1489–1502.

Gesler, W. 1986. The uses of spatial analysis in medical geography: a review. *Soc. Sci. Med.*, 23, 963–973.

Gouveia, D.S.A.; Nobre, F.F. 1991. The occurrence of meningococcal meningitis in a Southern region of Brazil, from 1974 to 1980, a point event model study. *Rev. Saúde Públ.*, 25, 103–111.

Haslett, J.; Bradley, R.; Craig, P.; Unwin, A. 1991. Dynamic graphics for exploring spatial data with application to locating global and local anomalies. *The American Statistician*, 45, 234–243.

Helfenstein, H. 1986. Box–Jenkins modelling of some viral infectious diseases. *Stat. Med.*, 5, 37–47.

Kafadar, K.; Stroup, D.F. 1992. Analysis of aberrations in public health surveillance data: estimating variances on correlated samples. *Stat. Med.*, 11, 1551–1568.

Marshal, R.J. 1991. A review of methods for the statistical analysis of spatial patterns of disease. *J.R. Statist. Soc. A*, vol. 154(3): 421–441.

Mayer, J.D. 1986. The role of spatial analysis and geographic data in the detection of disease causation. *Soc. Sci. Med.*, 17, 1213–1221.

Nobre, F.F.; Stroup D.F. 1994. A monitoring system to detect changes in public health surveillance data. *Int. J. Epidemiol.*, 23, 408–418.

Naus, J.I. 1965. The distribution of the size of the maximum cluster of points on a line. *J. Am. Stat.* Assoc., 60, 532–538.

Sanson, R.L.; Pfeiffer, D.U.; Morris, R.S. 1991. Geographic information systems: their application in animal disease control. *Rev. Sci. Tech. Off. Int. Epiz.*, 10, 179–195.

Sayers, B.McA.; Mansourina, B.G.; Phan Tan, T.; Bögel, K. 1977. A pattern analysis study of a wild-life rabies epizootic. *Med. Inform.*, 2, 11–34.

Scholden, H.J.; de Lepper, M.J.C. 1991. The benefits of the application of geographical information systems in public and environment health. *Raprr. Trimest. Statist. Sanit. Mond.*, 44, 160–170.

Selvin, S.; Merrill D.; Schulma, J.; Sacks, S.; Bedell, L.; Wong, L. 1988. Transformations of maps to investigate clusters of disease. *Soc. Sci. Med.*, 26, 215–221.

Splaine, M.; Lintott, A.P.; Angulo, J.J. 1974. On the use of contour maps in the analysis of spread of communicable disease. *J. Hyg.* Camb. , 74, 15–26.

Stroup, D.F.; Thacker, S.B.; Herdon, J.L. 1988. Application of multiple time series analysis to the estimation of pneumonia and influenza mortality by age 1962–1983. *Stat. Med.*, 7, 1045–1059.

Stroup, D.F.; Williamson, G.D.; Herndon, J.L. 1989. Detection of aberrations in the occurrence of notifiable diseases surveillance data. *Stat. Med.*, 8, 323–329.

Twigg L. 1990. Health based geographical information systems: Their potential examined in the light of existing data sources. *Soc. Sci. Med.*, 30, 143–155.

Walter, S.D. 1993. Visual and statistical assessment of spatial clustering in mapped data. *Stat. Med.*, 12, 1275–1291.

Watier L.; Richardson, S. 1991. A time series construction of an alert threshold with application to S. bovismorbificans in France. *Stat. Med.*, 10, 1493–1509.

Weinstock, M.A. 1981. A generalised scan statistic test for the detection of clusters. *Int. J. Epidemiol.*, 10, 289–293.

Zeng, G.; Thacker, S.B., Hu, Z.; Lai X.; Wu, G. 1988. An assessment of the use of Bayes' theorem for forecasting in public health: the case of epidemic meningitis in China. *Int. J. Epidemiol.*, 17, 673–679.

Case Studies from the South

Towards a Rural Information System

David le Sueur, Sipho Ngxongo, Maria Stuttaford, Brian Sharp, Rajendra Maharaj, Carrin Martin, and Dawn Brown[1]

Introduction

The findings presented in this paper are based on a project currently being carried out in South Africa entitled "Towards a Spatial Rural Health System." The information system being used for the project is termed the malaria information system (MIS), as it is based on the infrastructure provided by the Malaria Control Program. This paper will outline the control program and the use of global positioning systems (GPS) and geographical information systems (GIS) for malaria control, as well as the role of such information for the development of other programs for schools, clinics, electricity, water supply, and so on.

The malaria information system was established using the existing health infrastructure in a rural region of South Africa. This region has a high incidence of disease, a poor socioeconomic infrastructure, and one of the lowest per capita incomes in the country. It is also important to note that people do not live in villages, but in patriarchal homesteads usually separated by 50–500 m, depending on the terrain. This situation poses some unique issues related to malaria control.

Incidence of Malaria in South Africa

The annual malaria incidence for South Africa rarely exceeds 10,000 cases, with approximately 50% of cases occurring in KwaZulu/Natal Province. This relatively low incidence is directly attributable to the control program, which covers some 28,000 km^2 within the province. It is important to note the relative increase in cases over the past 10 years, which is largely attributable to factors such as agricultural development (Sharp 1990; Ngxongo 1994), vector drug resistance (Freese et al. 1988), population migration (cross border), and an alteration in malarial mosquito behaviour (Sharp and le Sueur 1990).

The historical perspective of the disease is also important, as it helps to contextualize the successes achieved by the program to date. In 1932, every magisterial district within KwaZulu/Natal Province reported cases of malaria;

[1]National Malaria Research Program, Medical Research Council. Sipho Ngxongo is with the Malaria Control Program, KwaZulu Health and Welfare.

estimated deaths over a 6-month period were in excess of 22,000 for a population at risk of 985,000 (le Sueur et al. 1993). The impact on the economy was enormous: construction of the Stanger Railway line ceased, and the sugar and tourist industries were crippled. Malaria is currently held at bay by extensive control efforts.

KwaZulu Malaria Control Efforts

Within the KwaZulu region, 52 malaria control areas have been established. For control purposes, the two northern districts of Ingwavuma and Ubombo have been subdivided into 27 malaria areas. Each of these is divided into 10 sections (Fig. 1). Within each section, every homestead is numbered by means of a malaria green card, which is stored in the eaves of one of the structures on the homestead.

The numbering system is updated every second year to account for population movement and growth. Several malaria control teams are stationed in the region, and are responsible for the annual application of a residual insecticide to the wall of every structure within one or more control areas. This insecticide application constitutes the major thrust against the mosquito vector.

In addition, the teams are responsible for carrying out active surveillance, whereby the population is routinely screened for infection. Infected cases are then treated and followed up to ensure parasite clearance. In this manner, the parasite reservoir is controlled, and transmission becomes limited.

This control infrastructure has been used to establish a computerised database which includes data on every homestead in the areas of malaria control (Table 1). The database is printed every second year and updated by control program staff during their annual spraying; changes recorded are subsequently made to the existing database. In the current update, information on Chief and Induna are being collected so that the database can be related to the tribal address system.

The database contains information on the head of the homestead, his "malaria home number", the type of structure, and the population living on the homestead. A set of GPS coordinates is subsequently added using a handheld GPS. Position is accurate to 50m, with a 95% confidence limit. A year was spent assessing the suitability of three different units in terms of accuracy, robustness, battery consumption, portability, and ease of use.

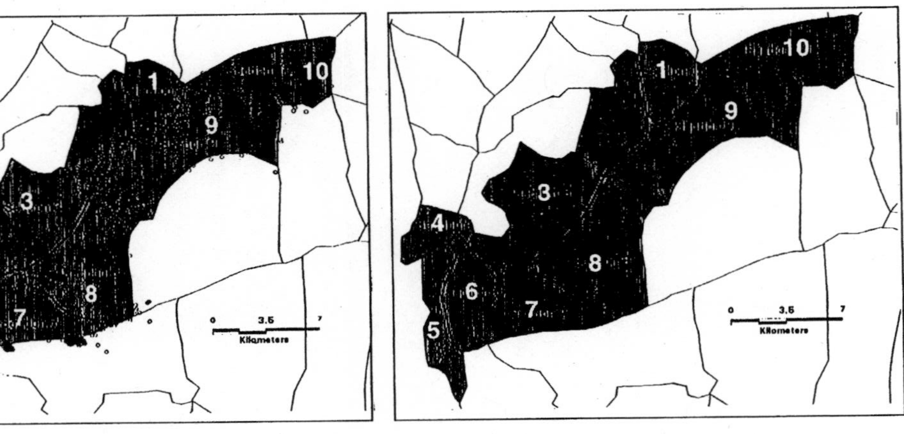

Owner	Area	COL2	COL2_2	Section	Houseno	Maltotal
MKATSHWA NXUMALO	MAMFENE	32.2434	-27.3659	1	1	0
SIZENI NGUBANE	MAMFENE	32.2426	-27.3661	1	2	0
TIMOTHY NGUBANE	MAMFENE	32.242	-27.367	1	3	0
BORNTXOGO SYAYA	MAMFENE	32.2393	-27.3678	1	4	0
MANDLAKAYISE GUMEDE	MAMFENE	32.2364	-27.3663	1	5	0
MLUNGISENI MOTHA	MAMFENE	32.2369	-27.3664	1	6	0
MALIWA MOTHA	MAMFENE	32.2371	-27.3656	1	7	0
MBONGISENI SIBIYA	MAMFENE	32.233	-27.3656	1	11	0
GEZUMUZI SBIYA	MAMFENE	32.2321	-27.3655	1	12	0
PHENEAS BUTHELEZI	MAMFENE	32.2332	-27.3664	1	13	0
DECEMBER MENYUKA	MAMFENE	32.2344	-27.366	1	14	0
ENOCK NTULI	MAMFENE	32.2355	-27.3678	1	15	0

National Malaria Research Programme, 1994

Fig. 1. Subdivision of malaria areas into sections. Each homestead within a section is numbered by a "malaria card" stored under the eaves.

Table 1. Homestead Database for Areas Under Malaria Control

Data Field		Example
Owner		Aaron Siyaya
Area		Mamfene
Section		8
Homestead Number		145
Population		12
Bed Bugs Present		Yes/No
Insecticide Used		DDT/FICAM/CYFLUTHRIN
Longitude		28° 12' 43 2" S
Latitude		30° 58' 39 7" E
Wall Surface	Mud	3
	Reed	1
	Wood	0
	Cement	0
	Painted	1

During the collection of homestead position data, information on clinic and school attendance was also collected. Table 2 shows the database format in which every malaria case is recorded. It is important to note that the malaria area, section, and house number are common to both databases, and act as a relational link between the two. It thus becomes possible to plot all cases at an area and section level; in areas where GPS coordinates have been added, it is possible to plot cases at the homestead level.

Research Findings

The implications of the data produced thus far are important, at both the "micro" (individual homestead) and "macro" level (geographic area such as the malaria area or section).

As outlined earlier, current control efforts have to a degree been compromised by factors such as drug resistance, changes in vector behaviour, agricultural development, and so on. It therefore became necessary to investigate supplementary control efforts.

Table 2. Malaria Case Incidence Database

Data Field	Example
Name	Velaphi Sithole
Case number	581
Month	January
Day	22
Notified district	Ubombo
Source district	same
Notified area	Mamfene
Source area	same
Section	8
Homestead number	145
Age	12
Sex	Male
Method of detection	Active/Passive
Physiological status	Healthy/Sick
Parasite	TGPF[a]

[a] Tropohzoite/Gametoc, vte/Plasmodium falciparum

This was the central theory on which this project was conceptualised. Le Sueur and Sharp (1988) have demonstrated the occurrence of a seasonal contraction and expansion of the malaria vector population. In winter the vector population is localized, and the larval cycle of the mosquito population increases from approximately 8 days in summer to 44 days in winter. As a result, the production of adult mosquitoes is reduced and the population is concentrated in the larval stages. If the winter sites could be located, they could be targeted for supplementary control measures, thereby enhancing overall control efforts. This was the central theory on which this project was conceptualised.

In instances when malaria cases can be pinpointed to specific coordinates, and when these coordinates can be plotted on a topographical map, control measures can often be easily determined. For example, it was discovered in the Mamfene Area that a prevalence of cases was occurring recurrently at a specific geographical location (Fig. 2). It was expected that a breeding focus probably existed at that location as well. Although little could be determined from first attempts to plot the findings at a 1:250 000 scale, when these coordinates were plotted on a 1:50,000 topographical map, they coincided with a water body in the area. Thus, it was determined that this location could be targeted for environmental management, to permanently alter or remove the breeding site. This action may or may not involve the provision of an alternative water source. The long term benefits would, however, be cost effective, when considered against the background of ongoing, long term control costs. Digitising of all the high risk areas at a scale of 1:50,000 is currently in progress. Should this level of detail be insufficient, then digitising of 1:10,000 ortho-photos will be carried out.

One example at the microlevel centres on the issue of agricultural development in malarious areas. Between 1976 and 1986, the Mamfene Area had an annual average of 12.6 cases of malaria. This increased to approximately 700 cases in 1987 as a result of excess water spillage due to agricultural irrigation. This spillage could have been avoided by proper planning in the development phase of the scheme, and highlights the need for intersectoral collaboration. The association between malaria cases and the scheme becomes evident when plotted onto a map: mosquito flight ranges and breeding sites were affected or created by the irrigation process, and almost all malaria cases which occurred in 1987 fall within the buffers of these new ranges/sites (Fig. 2). The cases which occurred along the canal were probably associated with breeding sites formed as a result of the interruption of natural water runoff during construction, as well as leakage from cracks in the canal.

At the macrolevel, a map indicating the annual incidence of malaria in Ingwavuma and Ubombo between 1980 and 1991 demonstrated that cases were not evenly distributed throughout the region, but were in fact highly concentrated (Fig. 3). It is likely that within these sections there will once again be a focus relating to the proximity of vector breeding sites. Such data can play an important role in the management of a malaria control program. These two northern districts generally contribute in excess of 75% of malaria cases occurring within the Province. Two important factors should be noted: the low incidence in the coastal region and the localization of high incidence to the Ndumu/Makhanisdrift area.

In the coastal regions (Fig. 4), the incidence of malaria is extremely low (< 0.2% per year), with many of the malaria sections reporting only one or two cases per year. The data has as yet not been divided into active and passive case detection. The almost negligible number of cases suggests that no local transmission is occurring; the cases reported are probably infected migrants entering the region.

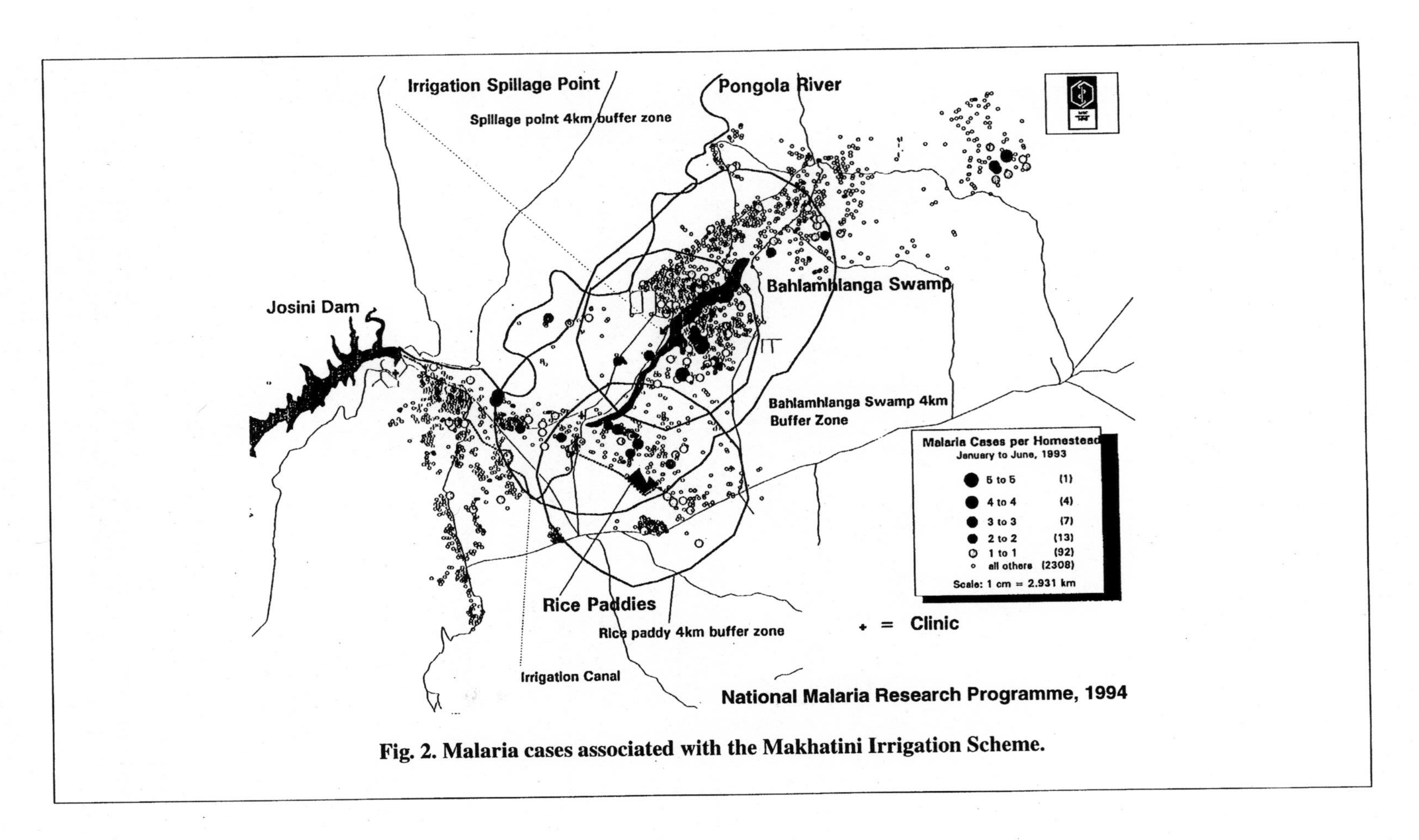

Fig. 2. Malaria cases associated with the Makhatini Irrigation Scheme.

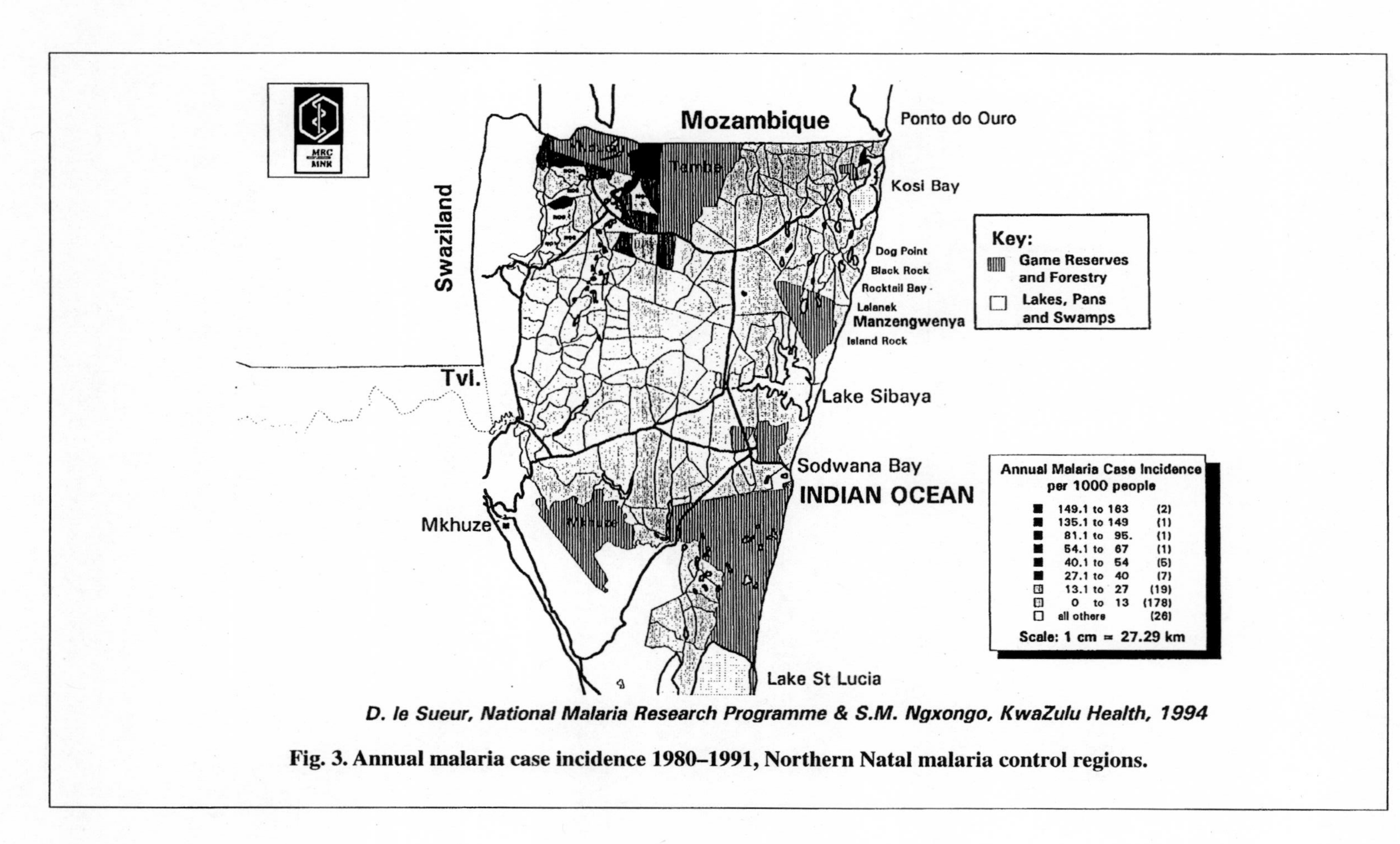

D. le Sueur, National Malaria Research Programme & S.M. Ngxongo, KwaZulu Health, 1994

Fig. 3. Annual malaria case incidence 1980–1991, Northern Natal malaria control regions.

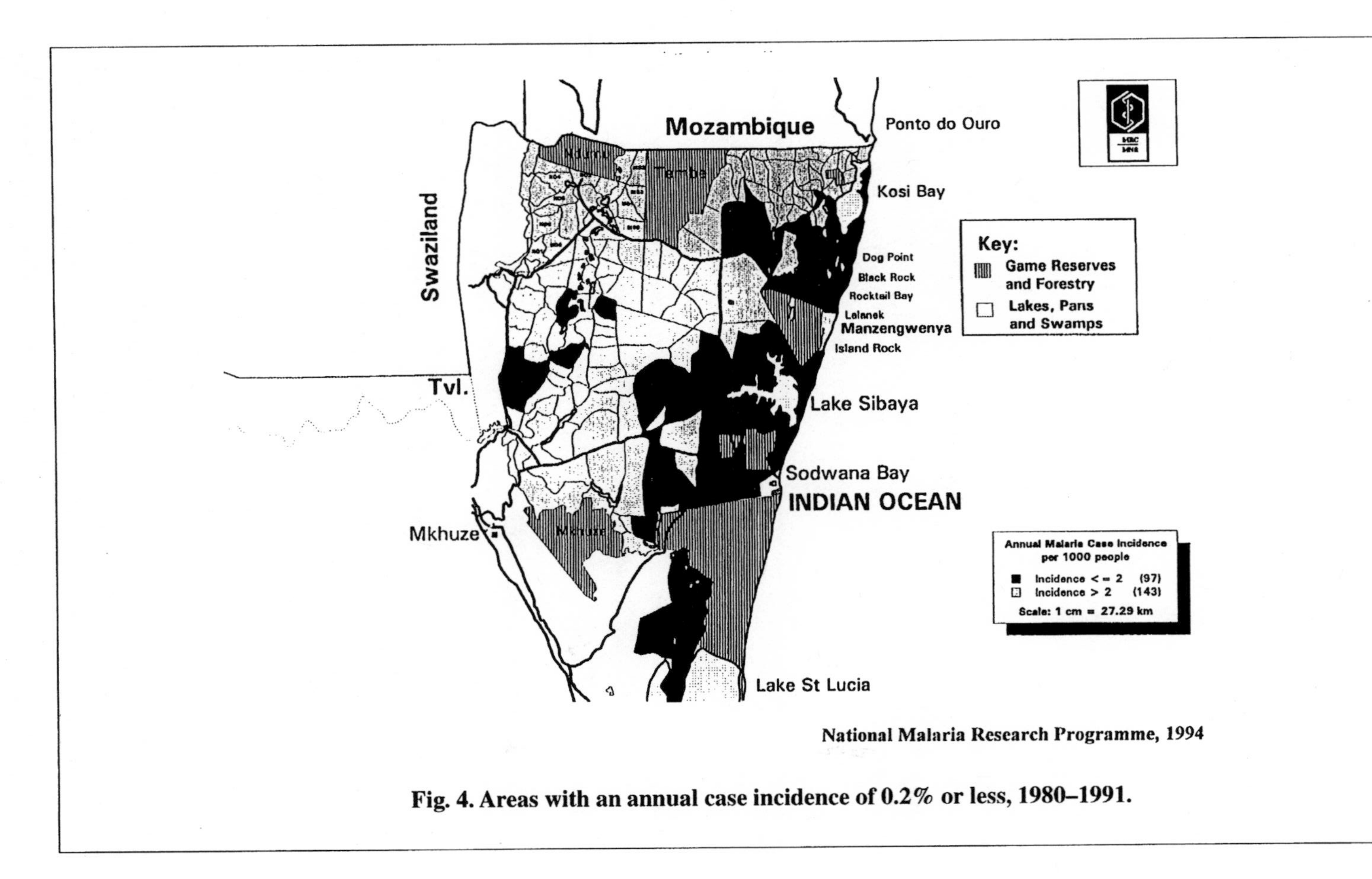

National Malaria Research Programme, 1994

Fig. 4. Areas with an annual case incidence of 0.2% or less, 1980–1991.

This argument is strengthened by extensive surveys that have been conducted in the region, in which almost no presence of vector mosquito species was found. The plotting of the actively detected cases will strengthen this argument. In 1938, the same area constituted the only region of continuous transmission in the country. This change can be largely attributed to the eradication of *Anopheles funestus* from the region and the fact that this vector's breeding sites are not utilised by the remaining vector, *Anopheles arabiensis*. Control measures do not currently exist in Mozambique, thus a need exists to continue with residual spraying to prevent the re-introduction of the former species into South Africa. Nevertheless, it can be concluded from the available data that the risk of future disease within this region is negligible. The malaria control management implications of this are discussed later.

The data does, however, have implications for tourists visiting the popular coastal region, with regard to the taking of prophylaxis. Due to chloroquine resistance within the region, tourists are required to take prophylaxis, which varies in cost between USD $25 and $45 per course. Residents in the area do not need to take prophylaxis. To facilitate dissemination of such information to the public, maps are published in the newspaper showing the transmission areas for the 4 weeks prior to publishing.

Mapping of Research Findings

Maps were created showing incidence data divided into two periods: 1980–1986 and 1987–1991. This division illustrates the recent increase in the number of cases within the area, which can be attributed to three factors:

- The advent of chloroquine resistance, first detected in 1985;
- Increased cross border population movement, which is highlighted by the extremely high number of cases occurring in the Mbangweni corridor between Ndumu Game Reserve and Thembe Elephant park, the main "corridor" for population movement to and from Mozambique; and,
- Agricultural development and the spillage of excess irrigation water into the veld (natural bush/grassland), which has altered the vector breeding patterns in the two southern areas. (These areas comprise the Makhatini Irrigation Scheme. There is also little doubt that the movement of infected, job seeking migrants to the scheme also plays an important role.)

House Construction Building materials used for house construction play a role in malaria transmission. The current insecticide of choice, DDT has a number of problems involved with its use, including high levels (30 x ADI) found in the breast milk of primiparous mothers, social resistance due to the bedbug and wall discolouration problems, environmental contamination, and malaria vector avoidance.

Many of the alternatives such as pyrethroids, however, are bio-degradable and thus not as stable as DDT. As a result, their lifespan (residual efficacy) is often limited on certain surfaces; thus consideration of the material of which house walls are constructed becomes important.

To be effective, these insecticides must have a residual effect for at least 5 months. In the laboratories of the project team, trials are currently being conducted on alternative insecticides with three different substrates.

The large number of nonmud substrates in the northern regions is largely a result of the sandy nature of the soil. Grass and reeds are therefore often used in construction, and generally provide a residual lifespan in excess of 9 months. This data can be used to target certain areas for cost effective, alternative insecticide applications.

Population Density Maps showing population density are also of use. Such data is currently being used to select a site in which bed nets (horizontal vector control) can be compared to residual spraying (vertical vector control) in terms of efficacy and cost-effectiveness. Similarly, this data can be used in conjunction with clinic catchment data to look at the potential for restructuring the parasite control component such that it can be carried out horizontally at the clinic/community health worker level.

Such investigations are essential to ensure that the restructuring of control efforts does not result in a loss of the enormous control gains made thus far, with the population within the area being placed at increased risk. This is especially important given the low levels of immunity in the population, in most areas, due to in excess of 40 years of control and consequent low levels of exposure.

The implications of such data to other sectors is obvious. There are approximately 100 malaria sectors in a district such as Ingwavuma. Data generated through the malaria information system is far more accurate than any other gathered to date, in relation to population distribution and density.

Maps on the population density/km^2, number of structures per sector, and distribution of infrastructure (schools, clinics, shops, tribal authority, and so on) are currently being prepared for the department of economic affairs for inclusion in a proposal to the Electricity Supply Commission. In some areas, distributions can be demonstrated at the individual homestead level.

This data would be extremely valuable for the provision of primary water supply in an area. Such provision is currently complicated in the Province due to the dispersed fashion in which the community lives. Thus, actual geographical distribution is important for the placement of such facilities/infrastructure and determination of the catchments they serve.

Other Uses for Maps Maps may be used to show all houses in the Ndumu area to give a quick, accurate picture of school catchments. Children within this area attend 14 different schools; not all are within the Ndumu malaria area. The majority of these are primary schools, with only a single high school. A map may, for example (Fig. 5a), show the homes of all pupils attending Ndumu High and St Philliphs (primary school).

Distance calculations (straight line) would show that the average pupil in the Ndumu area, attending Ndumu High, travels 4.7km each way to school, every day. One pupil lived 20.1km from the school. It is assumed that the pupil would be boarding with friends. This data has to be considered in the light of the poor transport infrastructure within the region, as well as the extremely low per capita income to pay for such transport.

By overlaying the MIS population data with catchment areas of local hospitals, it was also possible to get a measure of the actual population that the hospital was serving as opposed to what it was officially supposed to serve (Fig. 5b). Such data, when combined with population distribution/density, is useful in planning projected resource needs and the distribution/requirement for satellite clinic facilities.

The number of imported cases of malaria per annum could also be presented in map form. It was discovered that between 1980 and 1991, the two main entry points into the country were Mbangweni corridor (between Ndumu and Tembe Game Reserves) and the Musi area.

It is interesting to note that the high level of imported cases in the Ndumu area coincides with this being the highest transmission area for the country. This highlights the importance of infected migrants acting as parasite carriers when entering the country.

The fact that no such correlation occurs in KwaNgwanase Area supports the earlier conclusion that vectors are largely absent from this region and thus transmission from migrants to locals does not occur.

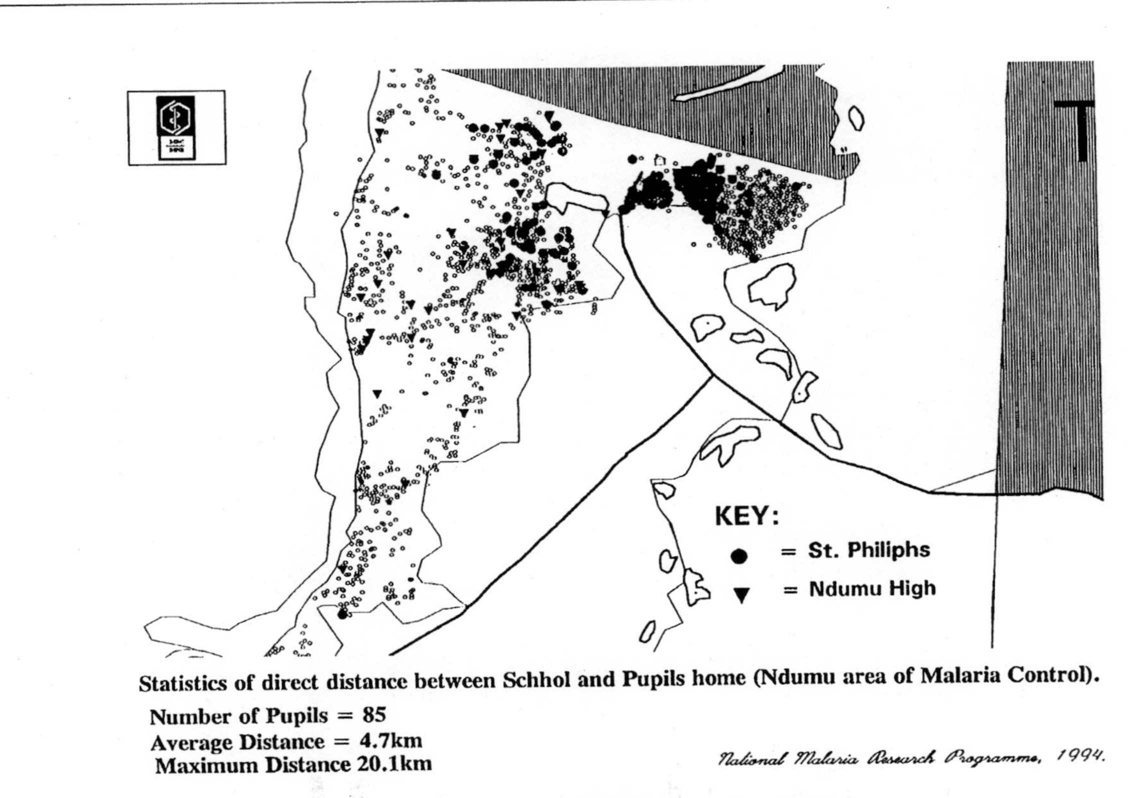

Fig. 5a. Determination of school/clinic catchments.

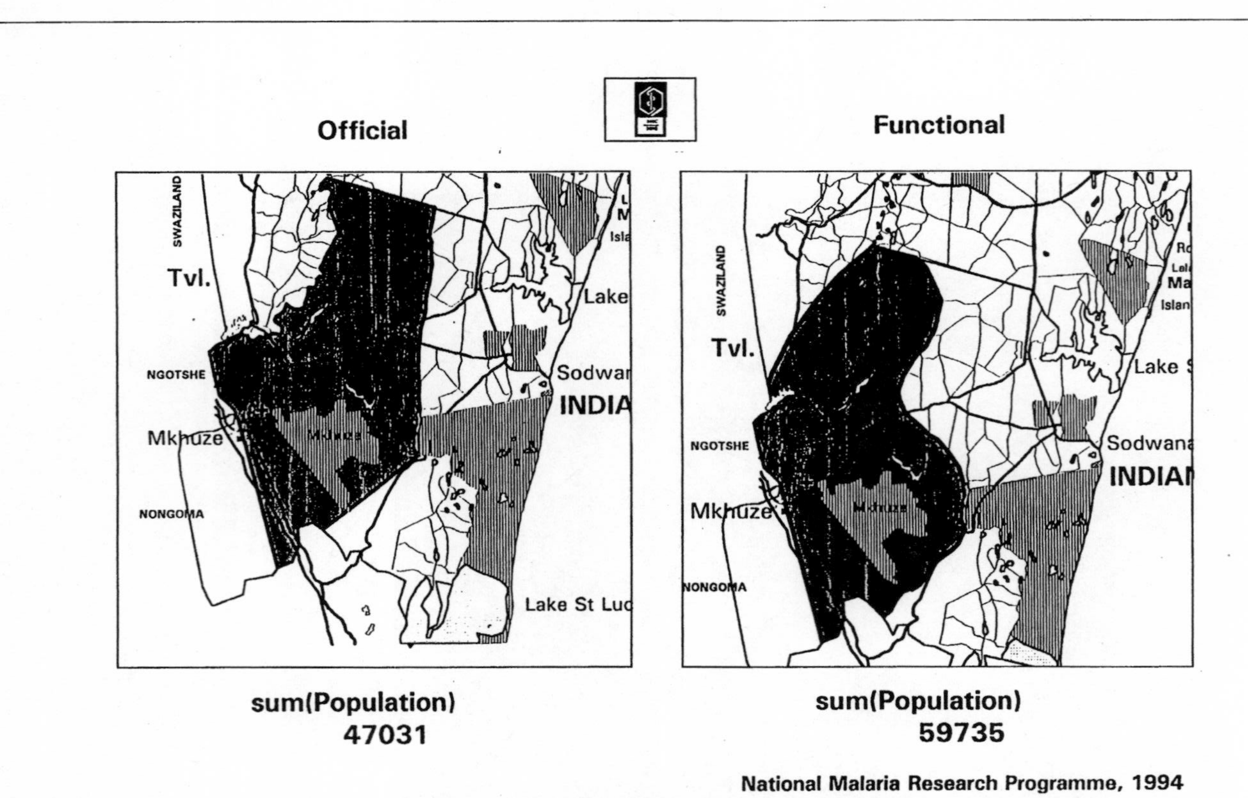

Fig. 5b. Catchment boundaries/population for Bethesda Hospital.

Implications of Research Findings

The foregoing data have numerous implications relating to the management of malaria in the country. Some of these are listed in the following.

Incidence of Malaria

The low incidence of malaria in coastal areas reflects very low transmission rates, and the high incidence in Ndumu, Makanis and other surrounding areas a high transmission rate. Malaria surveillance agents are currently deployed throughout malarious areas.

The project has shown that:

- More resources should be channelled to high risk areas, i.e., the need for redeployment of surveillance agents from low risk areas to high risk areas;
- Parasite surveillance staff members remaining in low risk areas will be required to do other primary health care functions; and,
- There is a need to continue with residual spraying in low risk areas.

Comparative Data 1980–1986 and 1987–1991

The above period shows a significant difference with a large increase of malaria incidence during 1987–1991. The factors responsible for this are:

- The intensification of the civil war in Mozambique which resulted in high population cross-border flow;
- The number of Mozambicans registered as Kwazulu/Natal pensioners who return monthly to receive their pension payments, resulting in increased cross border movement;
- The increase in vector drug resistance;
- The ideal climatic conditions for vector breeding;
- The commercial agents from non-malarious areas trading in malaria areas;
- The agricultural and other developments projects which disrupt and/or transform vector breeding sites, resulting in increased transmission; and,
- The recent establishment of two entry points along the Kwazulu/Natal-Mozambique border.

Conclusion

The conclusions that can be drawn from this project are:

- Malaria is not static;

- Malaria is a regional problem needing regional cooperation and uniform control measures in the southern area of South Africa;
- The need for intersectoral collaboration is highlighted by the fact that the government is spending $60,000 per annum for vector control to contain the malaria problem in and around the irrigation scheme. This does not include the cost of house residual spraying and the cost of treating the increased number of cases or their detection. This could have been avoided by consultation between health and agriculture specialists in the planning stages of the development scheme;
- There is a need to establish a parasite screening facility at each entry point to the country;
- There is a need to continually monitor the malaria situation. To facilitate this, the GIS has been disseminated to control program staff. This allows for a weekly production of maps based on notified cases, and highlights high risk focal points, enabling additional control measures such as larviciding to be applied. These maps will be used by management to facilitate decision-making regarding control activities at the weekly staff meetings; and,
- The establishment of such a "health based information system" using existing infrastructures can facilitate inter-sectoral collaboration. The information system built on the malaria control program (using the existing infrastructure and thus only utilizing an annual budget of $3,000 to establish the database) has wide implications throughout the region, for such sectors as education, water provision, sanitation, electrification, establishment of Primary Health Care, and so on. We believe that widespread usage of the information aids to offset the cost of malaria control, facilitates targeted development, and ensures that the system is sustainable.

Acknowledgments

The entire control program staff are thanked for their contribution to making this project possible. The Health System Trust is acknowledged for its financial support of this project, without its assistance, the project would not have progressed as rapidly as it has over the past six months. The Medical Research Council is acknowledged for baseline funding over the past 3 years.

References

Freese, J.; Sharp, B.L..; Ngxongo, S.M.; Markus, M. 1988. In vitro confirmation of chloroquine resistant Plasmodium falciparum malaria in KwaZulu. *South African Medical Journal*, 74, 576–578.

le Sueur, D.; Sharp, B.L. 1988. The breeding requirements of three members of the Anopheles gambiae Giles complex (Diptera: Culicidae), in the endemic malaria area of Natal, South Africa. Bulletin of Entomological Research, 78, 549–560.

le Sueur, D.; Sharp, B.L.; Appleton, C.C. 1993. A historical perspective of the malaria problem in Natal, with emphasis on the period 1928–1932. *South African Journal of Science*, 89, 232–239.

Ngxongo, S.M. 1994. The epidemiology of malaria in KwaZulu, 1980–1991. MSc thesis, University of Natal, Pietermaritzburg, Durban.

Sharp, B.L. 1990. Aspects of the Epidemiology of malaria in Natal Province, Republic of South Africa. PhD thesis. University of Natal, Durban.

Sharp, B.L..; le Sueur, D.; Bekker, P. 1990. Effect of DDT on survival and blood feeding success of Anopheles arabiensis in northern KwaZulu, Republic of South Africa. *Journal of the American Mosquito Control Association*, 6(2j), 197–202.

A GIS Approach to the Determination of Catchment Populations Around Local Health Facilities in Developing Countries

H. M. Oranga[1]

Introduction

Health care systems in Sub-Saharan Africa face increasingly diverse and complex health problems, rapidly growing populations, and severe resource constraints. Rational allocation of scarce resources is difficult, and is dependent on the size of catchment populations. Expensive hospital-based health care systems are protected by strong vested interests, reorientation is mainly rhetorical, and primary health care is making only slow progress.

Health care management and the use of health information at the local level are restrained by highly centralized decision-making processes. Several other weaknesses further restrict the usefulness of the health information system in Kenya. Some highly desirable information such as population-based epidemiology, service quality data and sociocultural information is not being collected.

Problems also exist in the flow of information from the field, including delays, nonreporting, nonresponse, and a generally unsatisfactory quality of generated data. Moreover, current reporting is largely restricted to acute and brief illness episodes in people fit and affluent enough to seek care at a health facility; those living far away, those too sick to travel, those worried about cost implications, and those chronically sick and disabled who have little to benefit from a visit, remain "invisible," neglected by the service system and overlooked by planners. Much of this is known by the local community, but unknown to the health care system and its staff.

The Kenyan population is heterogenous with ethnic, religious and socioeconomic differences influencing illness concepts and demands for health care. However, the only available information on the local population is the Census of Population, normally conducted every 10 years, which does not coincide with the actual catchment populations served by health facilities. The

---

[1]Geographic Information Systems (GIS) Unit, African Medical and Research Foundation (AMREF), Nairobi, Kenya.

health information system should reflect these circumstances, but available data are almost exclusively about care-seeking clients and their service utilization. A need exists for a more precise and complete description of the catchment population and health situation. It is important to generate this information at village, community and division levels.

Three interrelated problems are prominent. First, health care planning and programing are highly centralized and largely incremental in nature. Second, most staff are poorly prepared for management and planning responsibilities, particularly at and below the district level. Third, existing health information systems are unsatisfactory both as tools for planning and as a basis for local community development and action. The use of geographical information systems for determining catchment populations may help to alleviate some of these inherent problems.

Geographical Information Systems (GIS)

Applications

A geographic information system (GIS) is designed to work with spatially referenced data (Maxwell 1976; EPA 1987). In other words, a GIS has both attribute and spatial databases with specific capabilities for manipulating spatially-referenced data, as well as a set of operations for working with the attribute data; in a sense, a GIS manipulates spatial and nonspatial data.

The essential components of a GIS are the same as those of any other information system, namely:

- Data acquisition,
- Preprocessing,
- Data management,
- Data manipulation and analysis, and
- Product generation.

Data should be acquired through field surveys, and should include analyses of existing secondary data from maps, aerial photographs, reports and other documents. Data accuracy and completion must be ensured.

The main emphasis of data analysis should be on the production of maps of demographic data. This activity relies on the collection of point data from known locations, and the global positioning system (GPS) is be used to record precise locations where sample data are located. At the end of the data collection activity, an associated count data (point data) will exist for each sample location. In the collection of data, efforts should be made to locate those areas from which no participants are coming. This will aid in the interpolation, by ensuring that areas with zero count (no patients) are taken into consideration.

This point data will be input into the GIS. To facilitate data input, the data should be in ARC/INFO ungenerated point format, that is: *Value, X-coordinate, Y-coordinate*, with *value* being the observed count for the location identified by the *X*-, *Y*-coordinates.

The next step is to generate a points coverage (digital version of a map) in GIS and to build a correct topology with related feature attributes. Additional information can be added to the attribute table, for example counts by sex. This coverage will form the input to the triangulated irregular network (TIN) module.

Using the ARC/INFO TIN routine, a TIN coverage will be generated. From this coverage, an isoline coverage will be produced. The final coverage will show those areas of equal numbers through the use of contours.

The determination of health facility catchment areas is more involved, as it requires the active participation of the local communities. Homesteads/villages are identified through a unique numbering system. As patients visit a health facility, they will be required to also provide their homestead/village particulars. In this manner, for a maximum of about two months, the spatial coverage for a facility serving a typical village (about 600–1,000 people) can be identified, and the specific areas where patients originate mapped.

The frequency of visits will then be tabulated/graphed. Interpretation of this data must take into consideration a number of factors, such as transportation costs, condition of roads, availability of local medicinemen, and so on. These factors will affect the size of the population visiting a health facility, frequency of visits, types and conditions of illnesses involved, and so on.

For each health facility, a plot showing from where patients come in a given period will be produced in the form of a dot map. All patients visiting health facility *n* will also be recorded as *n, X-Coordinate, Y-Coordinate*, where *X*-, *Y*-Coordinate is the homestead location. Isoline maps produced from this data will show the geographical reach of health facility *n*. Areas of overlap between catchment areas will be determined by overlaying maps from health facility *n* and *n+1*.

Health planners are confronted with the problem of selecting the optimum location for a health facility in relation to the spatial distribution of the catchment population and infrastructural facilities already in place.

Although this is not an easy exercise, the application of GIS may aid in decision-making. A study to find the best location for a health facility would start by identifying homesteads, road networks and health facilities in the study area. For the allocation of health facilities, a number of variables such as road conditions/types and travel times will have to be incorporated into the database before running the allocation function. The various solutions generated will have to be evaluated against the criteria specified for siting the health facilities. Such criteria may be to locate a facility where travel costs are minimized.

Software Although there are several softwares on the market, the pc ARC/INFO system is preferred for its extra spatial analytical capabilities. However, it needs a number of modules to facilitate work in the project described above. These are:

- A TIN module, for generating triangulated irregular networks,
- A NETWORK module, for resource location/allocation and selecting the shortest path between a given set of points; and
- NEXPERT, an expert decision system (EDS).

The development of EDS with GIS for health management is technologically feasible. Determination of catchment populations in the nomadic communities is more cumbersome and complex. The only reliable source of information is remote sensing. It is in the above situations that remote sensing data, particularly from the AVHRR sensor on board the NOAA-8 satellite, would be recommended (Wetmore and Townsend 1975; Shlien 1977). Plans are underway to test the feasibility of the technology in mapping and sampling of the nomadic communities. However, a need exists to consolidate the knowledge base required for implementing a satisfactory system. The acquisition of the ERDAS system should enable satellite image processing to be performed. In addition, the ERDAS system can easily share data with the pc ARC/INFO system. A multitasking operating system is necessary for improving machine usage, which tends to be underutilized, especially during the slow processes of digitizing and topology building.

Conclusion

The application of the GIS technology for determining catchment populations is feasible. The data required from health facilities serving a typical African village would take a maximum of two months to collect. The incorporation of remote sensing is also needed for handling migratory populations, such as the nomadic communities found in the semi-arid and arid lands in eastern and northern Kenya.

References

Cohen J.M.; Hook R.M. 1987. Decentralized planning in Kenya. *Public Administration and Development,* 7, 77–93.

EPA. 1987. US cancer mortality rates and trends, 1950–1979 (EPA 600/1 -83/015). U.S. Environmental Protection Agency, Research Triangle Park, NC, USA.

GOK. 1984. District focus for rural development. Nairobi Office of the President, Nairobi, KE.

Mason, T.J.; McKay, F.W.; Hoover, R.; Blot, W.J.; Fraumeni, J.F. Jr. 1975. Atlas of cancer mortality for U.S. counties: 1950–1970. DHEW Publication No. (NIH) 75–780. Washington, D.C. Government Printing Office, WA, USA.

Maxwell, E.L. 1976. Multivariate system of analysis of multispecral imagery. *Photogrammetric Engineering and Remote Sensing,* September, pp. 22–24. Edmonton, Alberta, Canada.

Ministry of Health. 1986. National guidelines for the implementation of primary health care in Kenya. Nairobi, KE.

Shlien, S.; Smith 1977. A rapid method to generate special theme classification of LANDSAT imagery. *Remote Sensing of Environment,* 4, 67–77.

Wetmore, S.P.; Townsend, G.H. 1975. A geographical mode for storage identification, Analysis of Ecological data. KREMU Technical Report No. 1. Nairobi, KE.

GIS Management Tools for the Control of Tropical Diseases: Applications in Botswana, Senegal, and Morocco

Isabelle Nuttall,[1a] D.W. Rumisha,[b] T.R.K. Pilatwe,[b] H.I. Ali,[b] S.S. Mokgweetsinyana,[b] A.H. Sylla,[c] and I. Talla[d]

Introduction

Although the geographical approach has long been integrated into tropical diseases control programs (OMS 1965), the linkage between geography and health in general have only come to the fore in recent decades (Verhasselt 1993). For tropical disease control, these links are particularly relevant as refugee movements, the continuous population flux between rural and urban areas, and environmental changes influence the distribution of vectors, reservoir animals and the human population, and determine the transmission of diseases.

Technological progress has led to the emergence of new computerized analysis tools, known as geographical information systems (GIS), whose use in the health field vary from the simple automated mapping of epidemiological data (Pyle 1994), to the sophisticated analysis of satellite images which demonstrate vector/environment relationships (Hugh–Jones 1989; Malone 1992; Perry 1991; Rogers 1991). GIS have already been widely used in other sectors such as the management of natural resources, agriculture, and rural and urban planning (Rideout 1992).

The purpose of this article is to show how, on the basis of specific examples drawn from WHO's experience with the Ministries of Health in Botswana, Senegal and Morocco, it is possible to meet the decision-making needs of countries through the use of GIS within tropical diseases control programs. This review illustrates how geographical information systems facilitate the monitoring and management of control programs and open new avenues for intersectoral collaboration.

[1a]Division of Control of Tropical Diseases, WHO, Geneva, Switzerland.

[b]Community Health Services, Gaborone, Botswana.

[c]Chief of Division of Statistics, Senegal.

[d]Chief Medical Doctor, Health District of Richard Toll, Senegal.

The Needs

Accessing data from different sources at all levels of the health care system is a general challenge. Although a large quantity of data is collected either routinely or through special surveys, the information is generally synthesized at the level where it was collected, then transmitted to the next higher level where it is aggregated and transmitted further, until it becomes difficult to isolate the basic information. Available data are often presented in the form of tables or isolated figures, the reading of which is a laborious and time-consuming task and which does not permit easy decision-making (Sandiford 1992).

Specific needs for health services vary according to the level of decision-making. The district medical officers, who in developing countries frequently move from one post to another, must be able to acquire a thorough knowledge of their area of responsibility rapidly and easily on the basis of data in their possession.

They are the key partners of the national disease control services, and must have the capacity to express the needs of their districts and to determine its health priorities. They are responsible for the reliability of the data communicated by the clinics, and consequently have specific needs with regards to training, increased awareness, and motivation of the local health personnel for data collection.

The regional medical officers are located at the interface of the national and local services. They must have accurate and reliable data for planning. At this level, they are often the chief coordinators of research projects carried out by foreign institutions, providing the link between research and action.

Epidemiological surveillance is an essential need for tropical diseases control programs (malaria, schistosomiasis, onchocerciasis, and so on). These programs must be capable of constantly generating updated information and should be useful in guiding field operations: when and where to intervene, which would be the most effective interventions, whether an intervention is feasible with the limited resources available, and so on.

At the central level, the statistics and epidemiology departments are responsible for determining long-term trends, assessing specific risk factors, and integrating data that are not directly supplied by the health services to support planning and management. Reliable but still readily understandable information should be available to decision-makers.

The Tool: GIS

A GIS is a computer based system for inputting, storing, accessing, analyzing and presenting spatially referenced data from various sources in the form of maps. The GIS can, therefore, create a link between spatial data and their related descriptive information. A GIS is a combination of hardware (computers, digitizing table, scanner, GPS (global positioning system), plotter, printer, and so on) and specific software. The input and output of GIS is determined by the available databases and the technical skills of the staff operating the system.

GIS can be used not only for automatically producing maps, but it is unique in its capacity of integration and spatial analysis of multisource data: population, topography, hydrology, climate, vegetation (satellite pictures), access routes (roads and railways), public infrastructure (schools, main drinking water supply), and health infrastructure, including epidemiological data on diseases.

The GIS is often misperceived as a sophisticated technology, requiring satellite imagery that is often inaccessible to developing countries. In reality, GIS enclose a wide range of hardware and software covering a span of affordability and technical performance. This technological diversity offers great flexibility so that it can be implemented in most developing countries according to their needs.

Implementation of GIS for the Control of Tropical Diseases: Applications in Botswana, Senegal and Morocco

Since 1989, the WHO Division of Control of Tropical Diseases in Geneva has been interested in the application of GIS (Yoon 1994). A case study based on data from a schistosomiasis control program on Pemba Island (Zanzibar) (Savioli et al. 1989), analyzed how the distribution of staff and health services would need to be adjusted if a vertical program of schistosomiasis control were integrated into the health care system. This demonstration alerted several health ministries of the potential use for GIS to improve the planning and management of control of tropical diseases. Since early 1994, GIS for health are being implemented in Botswana and Senegal with support from IDRC, and a feasibility study has been conducted in Morocco.

Schistosomiasis Control in Botswana

The first infection of *Schistosoma haematobium* in Botswana occurred in Lobatse Railway in 1930. Transmission spread, and by 1978 *S. haematobium* infection was prevalent throughout the country at rates ranging from 0.6% to 14%. *S. mansoni* infection was first detected in Ngamiland District in 1965 with a prevalence rate of 3.1%. However, by 1983 a survey of primary school children in Maun revealed an 80.3% prevalence of *S. mansoni* and 1.4% of *S. haematobium*.

Recognizing the public health importance of schistosomiasis, a control program was established in 1985 in Ngamiland District with two main objectives.

- To develop a public health Schistosomiasis Control Program through a combined approach of mobile teams and the primary heath care system.
- To control *S. mansoni* infection by reducing prevalence by at least 75% and reducing heavy infections (> 100 eggs/gram of faeces) by at least 90% among school children by January 1988.

The control program was coordinated by a multidisciplinary National Task Force. The program sought multisectoral involvement to ensure sustained commitment to the goal. The terms of reference were:

- To develop a national plan of action for schistosomiasis control,
- To prepare a final project document for Ngamiland schistosomiasis control, and
- To carry out an annual review of operational and administrative aspects of the control program.

Local teams undertook case detection, treatment data collection, recording, and analysis. The Kato technique was used for diagnosis, and cases were treated with praziquantel 40mg/kg in single doses. Case detection occurred mainly through school and community surveys.

Simplified data collection and analysis procedures were used. The data collection forms were designed according to the objectives of the program and stored in DBF with the aid of a private NGO.

School Survey Results

Six school surveys were undertaken between 1986 and 1991. Of the 16,402 enroled children in 45 schools, 15,943 (97%) were examined during the first survey. Overall prevalence of *S. mansoni* was 28.7%, while prevalence of heavy infection was 4.1%. By 1989, the survey was carried out only in schools where the prevalence remained above 10%. Within these, overall prevalence of *S. mansoni* was 8.4%, while prevalence of heavy infection had fallen to 0.2%.

The sixth school survey in 1991 was conducted again where previous prevalences remained above 10%; 5,381 children were examined. Prevalence of heavy infestation had decreased to 0.01%, with total prevalence at 6.7%. Only one primary school had a much higher rate, 14.68%. By 1993, however, the prevalence rate in that school had dropped to 8.3%.

Community Survey Results

Mobile field teams carried out community surveys after consultation with the local authorities and local health facilities. These surveys were designed to identify and treat adults and children not attending schools (about 30% of the school age population) who were at risk of exposure to the disease. Two community surveys were conducted in the villages in Ngamiland in 1985–86 and 1987–88, respectively.

The prevalence of infection in all villages in Ngamiland was much lower than in the schools (19% during the first survey). Following selective chemotherapy of affected individual wards and villages, only two of the 25 wards in Maun, and one of the 12 villages in Ngamiland, showed an increase in prevalence during the second survey. One explanation for this might be the mobility of the population.

The crucial issue that remains to be addressed is monitoring the trend in prevalence rate reduction in relation to changing environmental factors. GIS was seen as an tool to assist this study.

Integration of Environment and Health with GIS in Senegal

Seeking to control the drought in Senegal, two dams were built, one 27km from the regional capital city of St Louis, the other in Mali territory, which lies some 1200km from St Louis. These dams have led to ecological changes that were responsible for a severe outbreak of schistosomiasis in Richard Toll in 1988.

The construction of the Diama dam was meant to stop salt water intrusion from the Atlantic Ocean into the Senegal River. Richard Toll is located 106km from St Louis. It is by far the largest rural agroindustrial city in the country. Its main industry is cane sugar manufacture, which attracts people from the surrounding areas where intestinal schistosomiasis is endemic.

The population is currently expanding so quickly that alimentary provisions such as clean water supply are lacking..The only source of water for domestic use remains the irrigation canals of the factory, and the River.

Schistosomiasis prevalence is estimated between 60–90% in Senegal, varying by region, and all age groups are equally affected. A recent study carried out in 1993 in the district demonstrated that the disease is spreading outward into other regions, in spite of the preliminary control measures that were undertaken.

The Ministry of Health currently recognizes that adequate control requires a deep and precise understanding of the distribution of the disease according to environmental data. This focus provides a relevant justification for using geographic information systems (GIS).

The Ministry of Health project has two main objectives.

- At the local level, to develop a better understanding of and policy to combat the proliferation of schistosomiasis infection through the use of GIS.

- At the national level, to utilize GIS as a tool for the integration of overall collected data (agricultural, environmental, and so on), and to facilitate decision-making about resource allocation.

At the present time, health data is being collected in five regional districts near St. Louis. An initial training course on GIS has also been completed.

Implementation of GIS

WHO's approach to the implementation of GIS in the control of tropical diseases is fundamentally different from that taken by universities or research laboratories. The latter, by virtue of their terms of reference, have the option of using sophisticated technological resources such as remote sensing satellite imagery and mathematical modelling. As WHO's approach is focused on the optimal use of resources already existing within a country, its process for implementing GIS for health follows six main lines:

- Identification of the needs and definition of the project area,
- Identification of the national GIS resources,
- Identification of the participants of the GIS project,
- Choice of the hardware and software,
- Definition of the required data, and
- Training.

Each country might see the utilisation of GIS in a different way, according to their needs and priorities. In Senegal, a severe epidemic of intestinal schistosomiasis due to *Schistosoma mansoni* (Talla 1990), which occurred in the Senegal Valley as a result of the Diama dam, highlighted the need for monitoring the course of this disease and others in space and their relationship to both environmental changes and health infrastructure. In Botswana, the development of national priorities for environmental protection identified the need to integrate the monitoring of tropical diseases within a national GIS. In Morocco, the epidemiological surveillance system needed to be improved to facilitate the elimination of schistosomiasis (Ministry of Public Health, Kingdom of Morocco 1993).

To ensure that the utilization of GIS is quickly undertaken, the system should be confined initially to a limited geographical area related to a priority disease/problem in the country. In Botswana and Senegal, starting from the expression of a specific need for schistosomiasis control, the GIS concept will be used in one region to integrate a whole set of data, a process which facilitates subsequent expansion while setting successive objectives.

It is necessary to identify the national GIS resources in other sectors. Specifically, the available boundary files, databases and technical skills should be determined. GIS is generally not used in the health field, while other sectors, such as the management of natural resources, agriculture and water resources, are already advanced in its application. Moreover, in some developing countries international agencies (e.g., UNITAR, UNDP, UNEP, FAO, World Bank) have helped to set up centres of excellence in GIS, cartography, and remote sensing. These centres have trained staff and geographical databases (automated maps), and form the nucleus of the national network of GIS users (Sahel and Sahara Observatory 1991).

In Botswana, the National Committee for Mapping and Remote Sensing, supported by a GIS users group, defined the standard formats to be followed in GIS. The group recommended that for each GIS project set up in Botswana, data should be compatible with the ARC/INFO format (Nkambwe 1994).

Since several departments of various ministries (e.g., Water Resources/Department of Water Affairs; Planning/Department of Town and Regional Planning; Agriculture/Department of Wildlife and National Parks) were already using GIS, the central government designated the Department of Surveys and Land to produce a digitized map of the country at a scale of 1:250,000. This map would comply with the defined standards and would be available and used by all the ministries concerned. At the same time, a private company with technical capabilities and special interest in nature conservation initiated a project to digitize a map of northern Botswana.

After identifying the available resources, the role of WHO, assisted by IDRC, was to support the initiatives in progress, both financially and through institutional collaboration. The mapping of Northern Botswana (1:250,000) was completed so that it could be used not only by the health structures, but by all ministries. An agreement was reached with the Department of Surveys and Lands that, after verification, the digitized cartographic database would be approved as an official map of Botswana. Similarly, training of the health staff involved in GIS was based on collaboration between the Department of Environmental Sciences of the University of Botswana, which provides appropriate training for the staff of certain ministries (e.g., Ministry of Local Government, Lands and Housing), a consultant of the ESRI conservation program attached to a private company, and WHO for the health component.

In Senegal, the Ecological Monitoring Centre (CSE) is the principal partner of the Ministry of Health. The CSE is a centre of excellence which was initially supported by the international agencies. It is now independent. Its terms of reference are the supply of reference data on national resources, the monitoring of indicators of environmental status, the management of a database integrated in a geographical information system, and the dissemination of information on the environment to planners and decision-makers. Within the GIS health project for the river valley, it is responsible for integrating the database of the Ministry of Water Resources (SIGRES project: Geographical Information System for the Management of Water Resources in Senegal) with all the other available data, and transferring them to the Mapinfo format for MacIntosh.

In Morocco, as in Botswana, there are two national committees working in the field of GIS: the national committee on mapping and the national committee on remote sensing. The task of these two committees is to coordinate activities between these two areas of study. This collaboration highlights the importance of developing GIS in other ministries (e.g., the Ministry of Mines and Geology, Water and Forestry, Water Resources, the State Under-Secretariat for the Environment, the Ministry of the Interior, the Royal Police Force). To cope with the demand for basic data, the Department of Land Conservation, Land Registry and Cartography has been entrusted with the task of digitizing all maps at a scale of 1:250,000. The automated maps will be available to any ministry requesting them, as is already the case with the paper maps. The Royal Centre for Spatial Remote Sensing, a public body working primarily in the field of remote sensing, serves as the national reference centre. It is particularly well endowed with regards to equipment, and it also provides training.

The participants in the GIS project are located at different levels within the health system. To ensure national commitment, the initiative to set up a GIS must be taken at the higher levels of the Ministry of Health. The decision-makers can then swiftly perceive the potential benefits of GIS and can participate in setting up the network of collaborators, which is one of the cornerstones for the success of the project. The Permanent Secretary of the Ministry of Health in Botswana, the Director of Public Health in Senegal, and the Director of the Health Programs in Morocco fully participate in the projects on GIS for health in their respective countries.

The practical implementation and monitoring of the project is specifically entrusted to a "GIS task force." The members belong to the division of statistics or epidemiology: epidemiologists, statisticians, computer scientists and geographers, according to the availability of local manpower. In Botswana, the core group is built around the unit of epidemiology, while in Senegal the leadership was taken by the director of health statistics. In Morocco, the team will be driven by geographers. These task forces are the preferred contact points of the national GIS centres of excellence. The local health services, in particular district chief medical officers, are the first suppliers of data to the system and the first users. In Botswana, two districts are integrated in the initial phase, in Senegal five.

When choosing the hardware and software to be used within the GIS projects, several considerations must be taken into account. Collaboration is stressed in this approach to the use of GIS for health in general, and for tropical diseases in particular. The GIS must remain a means of analyzing information and of making good use of what is available; it is neither necessary nor desirable for departments of statistics or epidemiology to turn into cartography units. Accordingly, preference is given in the initial stages to equipment for visualizing information rather than equipment for data input. For each project, three elements have been taken into account for the choice of software:

- The specific needs expressed by the health sector,
- The degree of progress reached by the country (other sectors) in GIS and the expertise of the potential partners, and
- The compatibility of the software selected with other softwares on the market.

A combination of these three criteria will often lead to a choice of software such as MapInfo, Atlas*GIS or ArcView.[2] These softwares permit simple mapping of the existing data and offer good compatibility with other softwares.

In Botswana, for example, the ARC/INFO format was fixed as standard for the government's GIS data. The complexity of using PC ARC/INFO led to the utilization of ArcView, which is much simpler to use. It can be used for visualizing geographical data directly in the ARC/INFO format and for attaching specific information to them. By choosing this software, it is possible to respect the high standards established in the country for GIS, while adapting to the needs expressed by the health sector. It ensures that during the initial phase of the project, the various participants (epidemiology unit, national program for control of parasitic diseases and the two district medical officers) work with the same tools. This offers the possibility, if necessary, of switching to more powerful softwares and collaborating with research laboratories which may wish to supplement the analyses with studies integrating satellite imagery.

Two types of data need to be included in the GIS: strictly geographical data and information relating to these data, known as "attributes." The geographical data include such items as administrative boundaries, the location of towns and villages, roads and railways, rivers and the main areas of vegetation. The attributes may be general (category of road or type of village), or more

---

[2]ARC/INFO, PC ARC/INFO and ArcView are registered trademarks of ESRI (Environmental Systems Research Institute Inc., Redlands, California, USA). MapInfo is a registered trademark of Mapping Information Systems Corporation (MapInfo Corp.), Troy, New York, USA. Atlas*GIS is a trademark of Strategic Mapping Inc.

specific (population at different dates, availability of a clinic, hospital, school, and so on). The epidemiological data are reported either for the village where the clinic is located or for the clinic's area of influence if it can be identified. These data are derived from specific surveys conducted previously or from routine reports.

The project in Botswana was initiated using data from exhaustive school surveys carried out by the Schistosomiasis Control Program between 1986 and 1991. In Senegal, routine data collected from health posts will serve as a basis for the analysis. In Morocco, surveillance data from the schistosomiasis control program will be integrated.

One of the key components for the utilization of GIS is the ability to ensure appropriate training. It has been demonstrated in Botswana and Senegal that training should be provided for everyone involved in a GIS project, and should be adjustable to the subsequent responsibilities of each person. A large part of the training should be devoted to the practical use of the tools. Decision-makers could be given shorter training, consisting of a general presentation of the concept of GIS and of the data to be incorporated. The potentials and limitations of GIS for decision-making should be stressed. It is desirable for decision-makers to be able to use the software selected so that they appreciate its usefulness and its limitations. Joint training for district medical officers and for members of the GIS task force ensures that everyone has the same basic knowledge.

Training is carried through national centres of excellence, using geographical and epidemiological data from the area of study. The involvement of these centres is essential. The presentation of concepts and the practical work refer more specifically to health problems, and it may be necessary to bring an external consultant for the initial training. Later, the members of the GIS task force will be able to take over the training.

Discussion

It is now accepted that the use of the computer in the health systems of developing countries "is a need, not a fashion" (Sepulveda et al. 1992). The ever-decreasing costs and the parallel increase in performance have greatly facilitated access to information technology during the last decade. Studies have shown that microcomputers not only lead to an improvement in the quality of decision-making and to more efficient and rational management of resources, but they also bring about a significant reduction in the cost of data-processing (Sandiford 1992). Geographical information systems benefit from the availability of computers, and their introduction into the management of tropical disease control programs is now feasible.

The constant simplification of the handling of GIS software may conceal both the importance of data in GIS and the limitations of GIS. For the control of tropical diseases, all applications begin at the periphery. If the data generated at the periphery has not be reliably collected and grounded, no level of sophistication of the technology at the central level can improve it.

The level of reliability of geographical data vary according to the scale of the map. A village, for example, appears as a dot on a map with a scale of 1:200,000 (1cm on the map represents 2km on the ground) and a clinic building cannot be individual identified. On a map at a scale of 1:50,000, a village occupies a whole area within which the health facilities may be visible. It seems that a basic map at 1:200,000 or 1:250,000 corresponds to the management needs of disease control programs conducted at country level because it makes it possible, through the enlargement and reduction functions (zoom in and zoom out), to obtain an overall view if necessary, but one that is still sufficiently detailed at district level.

It is necessary always to bear in mind the scale of the basic map, for even if the functions of the software make it possible to enlarge the maps almost infinitely, the precision of the maps is determined at the time of input (a pool 100m in diameter — or theoretically 0.5mm on the map — can never be individual identified at 1:200,000).

In a given country, the first ministry or agency which tends to use GIS at the national level with reliable geographical data must meet the cost of inputting these data. If the expenditure has been high, the ministry will tend to try to protect the use of this data like an asset. It is, therefore, necessary to create a clear definition of the procedures for information exchange. A financial contribution could be charged for the digitized maps, as is commonly done for paper maps. What is paid for is not the information in itself, access to which is unrestricted, but the work of inputting this information. Provided that the amount charged is reasonable, this seems one of the safest ways to avoid duplication of the work, while providing some return on the basic investment. Each agency can then supplement the basic map according to its own data, integrate information from different sources, and conduct its own analyses.

As opposed to research where the reliability of the health data, and especially those related to the diagnosis of a disease, are extremely important in the process of risk factor identification, control programs based on surveillance systems might benefit of the integration of routine data into a GIS for a better management. Frerichs (1991) states that information feedback is the very key to the success of surveillance systems, and is facilitated by the utilization of information technology. He stresses the importance of graphs in the presentation of results and refers to the value of mapping. GIS do in fact meet this need for feedback to the district level by enabling routinely collected data to be presented easily and attractively. This process is by no means new and merely replaces the conventional map displayed in the offices of district medical officers on which the health posts are represented by pins of various colours. Nevertheless, GIS provides an advantage, inasmuch as it facilitates what Sandiford calls the "ritualization" of the interpretation of routine data (Sandiford et al. 1992). It is not the fact that the nurse regularly takes the patient's temperature that will enable the surgeon to detect a postoperative infection, but the fact that the information is plotted on a chart and regularly examined at the patient's bed.

In Senegal, for example, by strengthening the management information system, the GIS will make it possible to analyze routine information. Sandiford (1992) states that people are generally somewhat hesitant to use routine data under the pretext that they are not reliable. Looking at this from the viewpoint of "action-led information systems," that is, systems in which it is not worth collecting the information unless it leads to a decision, it becomes apparent that the lack of quality of the data is often due to their underutilization. As soon as the information is used, the errors and anomalies are rapidly corrected through the feedback process.

This review indicates that GIS should form an integral part of surveillance systems, for it is one of the few tools meeting the need of monitoring the distribution of a disease in space. GIS surveillance of tropical diseases have been mentioned elsewhere. In Israel in 1992, the surveillance of imported malaria cases, combined with the identification of anopheles breeding sites, led to the precise identification of the intervention areas and thus enabled malaria transmission to be kept within bounds (Kitron 1994).

In South Africa (Le Sueur, personal communication) the cartographic representation of geographically referenced databases of the malaria control program makes it possible to locate the high-risk areas in space and time. In the onchocerciasis eradication program in Guatemala, GIS are used to identify communities at risk and as tools for assisting in ivermectin distribution (Richards 1993).

In Morocco, the same approach will enable the schistosomiasis control program to concentrate its efforts on areas where the disease is still present, while maintaining active surveillance throughout the country, to attain the objective of eradicating the disease. The WHO/UNICEF program for the eradication of guinea-worm disease also relies on the mapping of endemic villages (Clarke 1991).

As was mentioned in the foregoing, the simplification of the handling of GIS may conceal the importance of the data, and furthermore the significance and limitations of mapping. It may be very tempting to map just any variable. Experience has shown that even the basic epidemiological concepts such as prevalence tend to be forgotten when mapping is carried out (for example comparisons based on an absolute number of cases). The greatest risk when analyzing superimposed data is the deduction of causal relationships from mapping. It is important to stress that the simple visualization of data under no circumstances permits the conclusion of a cause-and-effect relationship between the various phenomena observed. Nevertheless, the GIS makes it possible in this context to demonstrate spatial relationships which might lead to subsequent in-depth epidemiological research (Scholten et al. 1991).

Given the current limitations of GIS for spatial epidemiological analysis, GIS still has an important role to play in the management of control of tropical diseases as it meets a real need in decision-making. When Frerichs (1991) claims that most developing countries are investing in administrators rather than in epidemiologists, and that the latter should therefore have the means to present rapid and readily understandable results to the administrators, it seems clear that the use of GIS meets a real need in decision-making.

The time of decision-makers is becoming increasingly valuable and they cannot afford to waste it by wading through pages of reports in order to extract the substance. They need brief presentations, focused on pertinent comparisons. When such reports are available, they often present conventional comparisons, such as the value of an indicator in terms of a theoretical reference, often defined at the central level, or a comparison of a variable from one year to the next. It is often more efficient in terms of management to make comparisons between similar health care structures or neighbouring regions (Sandiford 1992). There are two objections to such comparisons: first, the level of aggregation of the data, which has already been discussed in the foregoing, and second, the difficulty of sharing the information. The very basis of the GIS can remove these two obstacles and provide a unique opportunity for meeting the needs of the decision-makers.

Acknowledgments

The use of GIS for the control of tropical diseases in Botswana and Senegal is financed by the International Development Research Centre, Ottawa, Canada. The author is grateful to ESRI (Environmental Systems Research Institute Inc., Redlands, California) for their support with the software PC ARC/INFO and ArcView. The author wishes to express her gratitude to Dr K.E. Mott for his comments on the draft of this article.

References

Clarke, K.C.; Osleeb, J.P.; Sherry, J.M. et al. 1991. The use of remote sensing and geographic information systems in UNICEF's dracunculiasis (Guinea worm) eradication effort. *Prevent. Vet. Med.*, 11, 229–235

Frerichs, R.R. 1991. Epidemiologic surveillance in developing countries. *Annu. Rev. Publ. Health*, 12, 257–280.

Hugh-Jones, M. 1989. Applications of Remote Sensing to the Identification of the Habitats of Parasites and Disease Vectors. *Parasitology Today*, 5(8), 244–251.

Kitron, U.; Pener, H.; Costin, C. et al. 1994. Geographic Information System in malaria surveillance: Mosquito breeding and imported cases in Israel, 1992, *Am. J. Trop. Med. Hyg.*, 50(5), 550–556

Malone, J.B.; Fehler, D.P.; Loyacano, A.F.; Zukowski, S.H. 1992. Use of LANDSAT MSS Imagery and Soil Type in Geographic Information System to Assess Site-Specific Risk of Fasciolasis on Red River Basin Farms in Louisiana, reprinted from *Tropical Veterinary Medicine*: Current Issues and Perspectives, Vol 635, *Annals of the New York Academy of Sciences*, 389–397.

Malone, J.B.; Huh, O.K.; Fehler, D.P. et al. 1994. Temperature data from satellite imagery and the distribution of schistosomiasis in Egypt. *Am. J. Trop. Hyg.*, 50(6), 714–722

Ministère de la Santé Publique, Royaume du Maroc. 1993. Processus d'élimination de la schistosomiase au Maroc, compte rendu de réunion. Morocco.

Mott, K.E.; Nuttall, I.; Desjeux, P.; Cattand, P. 1994. New geographical approaches to control of tropical diseases: some parasitic zoonoses. Proceedings of a WHO consultation on geographical methods in the epidemiology of zoonoses, Wusterhausen (Germany) 30 May – 2 June 1994. WHO, Geneva, Switzerland.

Nkambwe, M. 1994. Geographical Information Systems (GIS) in Botswana: introducing a new technology for environmental information management. (Unpublished)

Nuttall, I. 1993. GIS for monitoring and control of tropical diseases. Demonstration diskette for PC.

OMS. 1965. La reconnaissance géographique au service des programs d'éradication du paludisme, Organisation Mondiale de la Santé, Division de l'éradication du paludisme, Genève, PA/264.65.

Perry B.D.; Kruska, R.; Lessard, P.; Norval, R.A.I.; Kundert, K. 1991. Geographic information systems for the development of tick-borne disease control strategies in Africa. *Prevent. Vet. Med.*, 11, 261–268.

Pyle, G.F. 1994. Mapping tuberculosis in the Carolinas, *Sistema Terra*, 3(1), 22–23.

Richards F.O. 1993. Use of Geographic Information Systems in Control Programs for Onchocerciasis in Guatemala, Bulletin of the Pan American Health Organization, 27, 52–55.

Rideout, T.W., 1992. Geographical information systems and urban and rural planning. Rideout, T.W., ed.. The Planning and Environment study Group of the Institute of British Geographers, 147 pp.

Rogers, D.J.; Randolph, S.E. 1991. Mortality rates and population density of tsetse flies correlated with satellite imagery. *Nature*, 351, 739–741

Sahara and Sahel Observatory/UNITAR. 1991. Survey of Observation Structures. UNSO.

Sandiford P.; Annett H.; Cibulskis R. 1992. What can information systems do for primary health care? An international perspective. *Soc. Sci. Med.*, 34(10), 1077–1087.

Scholten, H.J.; de Lepper, M.J.C. 1991. The benefits of the applications of Geographical Information Systems in public and environmental health. *Wld Hlth Statist. Quart.*, 44 (3), 160–170.

Sepulveda, J.; Lopez-Cervantes M.; Frenk J. et al. 1992. Key issues in public health surveillance for the 1990's. International Symposium on Public Health Surveillance, Atlanta. April, Atlanta, GA, USA.

Talla, I. et al. 1990. Outbreak of intestinal schistosomiasis in the Senegal River Bassin. *Ann. Soc. belge Med. Trop.*, 70.

Verhasselt, Y. 1993. Geography of health: some trends and perspectives, *Soc. Sci. Med.*, 136(2), 119–123.

Yoon S, 1994. The value of geographic information systems with regard to schistosomiasis control. In de Lepper, M.J.C.; Scholten, H.J.; Stern, R.M.; Kluwer, R. eds. The added value of geographical information systems in public and environmental health. Dordrecht: Academic Publishers and World Health Organization, Regional Office for Europe, Copenhagen. (In press)

The Use of Low-Cost Remote Sensing and GIS for Identifying and Monitoring the Environmental Factors Associated with Vector-Borne Disease Transmission

S.J. Connor,[1] M.C. Thomson,[1] S. Flasse,[2] and J.B. Williams[2]

Introduction

Satellite Imagery

The past few years have seen a rapid growth in the number and capabilities of earth observation satellites, which can be used for environmental monitoring and hazard assessment (Wadge 1994). Images from these satellites can be used to survey large regions which are difficult to access on the ground, and to monitor changes in the distribution of natural resources and variations in weather systems. However, despite the increasing availability of such images, the products of remote sensing remain largely underutilized, particularly in less-developed countries.

Vector-Borne Disease Monitoring

While some remarkable successes have been achieved in the battle against vector-borne diseases, such as the Onchocerciasis Control Program in West Africa, the threats to health from vector-borne parasitic diseases are at least as important now as at any time in history. This is illustrated by the following two examples.

The incidence of visceral leishmaniasis has risen worldwide, and southern Sudan is currently experiencing a major epidemic (WHO 1993). The disease's sand-fly vector is associated with a particular woodland type. Changes in the distribution of tree cover in Sudan as a result of ecological and/or developmental processes may have important consequences for the spread of this disease.

Malaria remains one of the greatest killers of young children in the developing world. It also imposes a considerable economic burden resulting from high morbidity levels within the adult population. While statistics on the scale of the disease vary considerably, it is inevitably the poorer sections of the poorest

[1]Liverpool School of Tropical Medicine, Liverpool, UK.

[2]Natural Resources Insitute, Chatham, UK.

nations which suffer most. The World Health Organisation estimates some 110 million clinical cases of malaria worldwide per year, over 80% of which occur in Africa south of the Sahara. Malaria is transmitted by anopheline mosquitoes, which breed in surface water pools where environmental conditions are suitable for both vector and parasite development.

Both spatial and temporal changes in environmental conditions may be important determinants of vector-borne disease transmission. Capable of identifying these changes, satellite imagery may be able to define and predict areas and periods of high transmission. Thus the potential for using remotely sensed images for monitoring and evaluating the factors associated with tropical diseases clearly merits further
investigation and research.

In 1985, the National Aeronautics and Space Administration (NASA) initiated the Biospheric Monitoring and Disease Prediction Project, the aim of which was to determine if remotely-sensed data could be used to identify and monitor environmental factors which influence malaria vector populations. Initial studies supported by this project used high resolution images from the LANDSAT[3] TM sensor to monitor the development of canopy cover in Californian rice fields. Changes in rice canopy cover over the season were successfully used to predict fields with high or low mosquito densities (Wood et al. 1991).

Unfortunately, the use of high resolution satellite imagery is limited by its cost (some US$4,000–5,000 per image at a scale of 60–180 km) and temporal resolution (16 days per repeat for the LANDSAT, and 26 days for SPOT).[4] Poor temporal resolution is a serious limitation as cloud cover may obscure a "scene" (area of the image) frequently enough as to allow only one good image of the region per season. While high resolution images may play an important role in producing baseline land use maps, their high cost and infrequent occurrence limit their use for large-scale temporal monitoring of vector habitats.

Image availability is an important requirement for the use of satellite data in monitoring the environmental variables associated with fluctuations in vector populations. Images must be obtainable with sufficient frequency to allow for comparison with changing vector bionomics, within a biologically meaningful time frame. This means that the images should be available at a cost low enough to make it possible to use a large number of sequential images over the area of interest.

---

[3]American commercial earth observation satellite.

[4]French commercial earth observation satellite.

The lower spatial resolution images from the geostationary and polar orbiting satellites (METEOSAT[5] and NOAA[6]) have been useful for a variety of relevant applications. These include famine early warning systems (FAO 1990), monitoring changes in rainfall conditions and vegetation associated with desert locust swarms (Heilkema and van Herwaarden 1993), assessing deforestation in tropical forests (Malingreau 1991), identifying environmental conditions known to favour tsetse fly reproduction (Rogers and Randolph 1991), and detecting flooding of the breeding sites of Rift Valley Fever vectors (Linthicum et al. 1990). Therefore, any disadvantage of the lower spatial resolution of these satellites may be offset by the benefits of their frequency of observation, low cost, and ease of availability.

Images as Information

Even where satellite images are readily accessible, the data cannot be used immediately. It must first be analyzed to produce information, which in turn must be presented in such a way to be able to influence decision making. The most appropriate facility for achieving this is perhaps a geographical information system (GIS), which is able to read, process, analyze, and present spatially-related data for effective interpretation and use for a variety of environmental and resource management purposes. Within the context of tropical vector-borne disease forecasting and control, such a system would enable the presentation of temporal and spatial dynamics of the disease, in a meaningful way, to planners responsible for national and regional control strategies. Real-time information relevant to potential surges in disease transmission could be included, enabling the initiation of rapid response strategies.

Accessing Low-Cost Imagery

Famine Early Warning System (FEWS)

It is possible to obtain some higher spatial resolution LANDSAT images for certain parts of the world through the Famine Early Warning System (FEWS). Where these images are available, they can often be obtained for the cost of reproduction from the EROS Data Centre.[7]

---

[5]European meterological satellite.

[6]National Oceanic and Atmospheric Organization.

[7]Earth Resources Observation Systems (EROS), Sioux Falls, South Dakota, 57198, USA.

Africa Real Time Environmental Monitoring Information Systems (ARTEMIS)

One very valuable source of satellite imagery has recently become available through the Food and Agricultural Organisation's (FAO) ARTEMIS program. This program has compiled an archive data set of NOAA-AVHRR NDVI images for the African continent for the period 1981–1991. These NDVI (Normalised Differentiated Vegetation Index) images are the product of the red and near infra-red sensors aboard the satellite, and indicate the amount of photosynthetic activity taking place in the image area with an initial resolution of 1.1km. The NOAA satellites pass over the same scene twice daily but have limited data storage available on board. They use the peak NDVI value to create an "optimal" composite image, known as global area coverage (GAC), every 10 days, which is then downloaded to the NOAA database. It has been suggested that this largely removes the problem of cloud cover over a scene.

The data is transformed from GAC (in a process including further subsampling, cloud clearing, and reprojection) into a 10-day maximum value composite (MVC), which minimizes atmospheric interference, sun angle, and viewing geometry effects (Holben 1986). It then has a resolution of 7.6km on the Hammer-Aitoff map projection.

It has been difficult to obtain reliable NOAA images, for certain areas, following the eruption of Mount Pinatubo in June 1991. This eruption produced tons of aerosol dusts which have seriously affected the transmission properties of the upper atmosphere. Workers at NASA have been busy correcting the data, which will become available in due course.

The NOAA imagery from ARTEMIS, while of spatially coarse resolution, is very useful in terms of its low cost (currently free) and high temporal resolution, allowing for seasonal monitoring of vegetation growth. The ARTEMIS data set is available from FAO's Remote Sensing Centre[8] on a CD-ROM, and comes complete with both Windows and PC-DOS software, allowing graphic display, preliminary analysis, and subextraction of the data set.

Local Application of Remote Sensing Techniques (LARST)

The LARST program was developed by the Natural Resources Institute of the Overseas Development Administration[9] (U.K.) in collaboration with other scientific groups. The aim of this program is to promote improved management of natural resources through cost-effective remote sensing. LARST develops

---

[8]Remote Sensing Centre, FAO, Via delle Terme di Caracalla, I 001000 Rome, Italy.

[9]Natural Resources Institute, Chatham Maritime, Chatham, Kent, UK ME4 4TB.

robust, low cost systems for local reception, processing, and use of satellite data. Within developing countries, it can be used to inform and assist with practical decision making (Williams and Rosenberg 1993; Sear et al. 1993;). LARST is currently active in several countries, such as Algeria, Ethiopia, Kenya, Indonesia, Namibia, Tanzania, Zambia, and Zimbabwe. These countries are exploring new ways of converting data from the METEOSAT and NOAA satellites to provide useful environmental information for such areas as agriculture, fisheries, forestry, water resources, and pest and vector control.

The LARST program is particularly valuable for those interested in using low-cost satellite data to monitor vector habitats, as it helps to provide real-time local-access NOAA NDVI imagery at full spatial and temporal resolution. LARST is able to do this by installing a ground receiving station within range of the scene of interest, enabling direct reception of the full satellite data stream. The local area coverage (LAC) images have a spatial resolution of 1.1km directly beneath the satellite, and can usually be received at least twice daily (the size of the image pixel is not constant since it increases as the viewing angle increases). The data received from the satellite can then be converted to whichever specific mapping projection is required, a process which involves resampling the data, usually to an image pixel size of 1km.

It may be possible to set up a LARST system "in country" or in a neighbouring country, through the national meteorological service. Where these facilities are not yet available, it may be possible to put together a case for establishing a minimal operation through donor assistance. The Overseas Development Administration (U.K.) has to date supported a number of these systems in sub-Saharan Africa.

Rainfall and Vegetation Monitoring

The NOAA–AVHRR NDVI images have been shown to be highly correlated with the Cold Cloud Duration (CCD) images from METEOSAT. The combination of these two images provides a good, proxy indicator of effective rainfall in areas receiving less than 1000mm rainfall each year (Dugdale and Milford 1988; Bonifacio et al. 1993).

They are particularly useful where ground-based meteorological stations are few and/or poorly dispersed. The availability of such images would be very useful for seasonal studies of the environmental factors associated with vector-borne diseases, such as rainfall, river flooding, changes in the water table, vegetation growth, and the location of vector breeding sites.

Geographical Information Systems

Geographical information systems (GIS) are defined here as computer-based systems for entering, storing, analyzing, and displaying digital geo-referenced data sets. The problems arising from the use and development of geographical information systems in developing countries, and some advice on their solutions, are outlined by Hastings and Clark (1991), in a special issue of *The International Journal of Geographical Information Systems* (Vol. 5), which is devoted to the consideration of GIS within a developmental context. Rajan (1991) has, with the support of the Asian Development Bank, produced a valuable overview of the use of remote sensing in GIS. The document includes a useful section covering some of the key issues and problems surrounding their use, such as: increasing commercialisation, human resources, institutional capacity, training, and technology transfer. A consultation report to FAO on the use of GIS in strengthening information systems for veterinary services in developing countries offers a comprehensive outline of hardware requirements and software considerations (Perry and Kruska 1993).

At present, GIS are seen primarily as research tools in the field of vector-borne disease; indeed, they will become an increasingly important research tool as geographical databases, models, and analysis procedures continue to develop at a rapid pace. However, the use of GIS in aiding decision support has also become an area of growing interest. Spatially referenced, interactive models have been developed to simulate the broader effects of development policy in Senegal (Engelen et al. 1992; Connor and Allen 1994). A system for the national control of foot and mouth disease in New Zealand (Morris et al. 1993) is probably the best example of how geographical databases and disease epidemiology models can be integrated into a decision support system. GIS designers are also beginning to provide analytical decision support tools as part of their options, and such systems are being promoted for a more participatory planning process (Eastman et al., in press; Hutchinson and Toldano, in press).

Of the many GIS available, perhaps the most obvious low-cost choice for the study of the environmental factors associated with vector-borne disease transmission is the raster-based system developed by the IDRISI Project at The Clark Laboratories for Cartographic Technology and Geographic Analysis[10]. The IDRISI Project is run on a non-profit basis and maintains close developmental links with the United Nations Institute for Training and Research (UNITAR) and the United Nations Environment Program's Global Resource Information Database (UNEP/GRID). Continuity and compatibility of resource formats shared between

---

[10]The IDRISI Project, Clark University, Worcester, MA, 01610-1477, USA.

different databases is therefore possible. Despite having a high degree of functionality, the software is inexpensive and easy to learn, and consequently has found much use in research, teaching, and training establishments worldwide. The project has a continuing commitment to provide a low cost, professional-level, modular GIS that is regularly updated in response to user needs, and which enjoys good technical support and a user network. The strengths and weaknesses of the IDRISI software for the following studies are currently being reviewed.

Case Studies

The case studies of our immediate attention are those related to the distribution of a specific type of woodland thought to be associated with the visceral leishmaniasis epidemic in southern Sudan, and the interregional and interyear variation in malaria prevalence throughout The Gambia. The use of GIS/RS in development program design and management is also important. It can help minimize the risk of creating unmanaged vector habitats, which may occur through development projects such as dams and irrigation schemes.

The "Eco-Epidemiology" of Leishmaniasis in Southern Sudan

The Nuer and Dinka people of southern Sudan are currently experiencing what is thought to be the worst epidemic of visceral leishmaniasis (kala azar) in history. Environmental and demographic changes and the problems associated with regional conflict exacerbate the conditions under which this disease spreads, and there is probably nowhere that this is more apparent. The epidemic began in 1984; the most recent reports have put the death toll at some 200,000.

The sandfly vector of kala azar in southern Sudan is known to be associated with mature *Acacia-Balanites* woodland (Quate 1964; Schorscher 1991). Howell et al. (1988) found that vegetation maps from 1952, of an area close to the epidemic area, showed a much more extensive distribution of forests than later maps produced in 1980. This loss of woodland was attributed to extensive and prolonged local flooding during the early 1960s, caused by rainfall anomalies in Kenya and Uganda. During a field visit by staff from the Liverpool School of Tropical Medicine, the relatively uniform and young age of the present woodland was noted. It has been suggested that the current epidemic of kala azar is, in part, associated with the recent local maturation of *Acacia-Balanites* woodlands (Ashford and Thomson 1991).

Operational medical services in the epidemic area require maps showing the current distribution of *Acacia-Balanites* woodland throughout southern Sudan. These maps are needed to investigate the link between the *Acacia-Balanites* woodland and known occurrences of the kala azar vector and parasite. Preliminary explorations of the ARTEMIS images from the area, as well as the 1km NOAA–AVHRR images obtained through the LARST facility in Ethiopia, have shown clear, delineated features believed to be associated with wooded areas observed on the ground.

It is hoped that these features can be identified and defined within the GIS and used as "training sites," which would allow the computer to extrapolate their distributions on a wider scale. Maps produced in this way could then be compared with historical vegetation maps of the same area and changes in canopy cover quantified. The maps could also be used, along with other epidemiological information, to help predict the spread of the current epidemic and to indicate where resources might be best mobilized (Connor and Thomson 1994).

Malaria Transmission Dynamics in The Gambia

The value of LAC NDVI data in explaining spatial variation in malaria transmission in The Gambia is currently being evaluated. The results of an extensive epidemiological and entomological survey in 1991 showed that malaria prevalence and intensity varied considerably from area to area, with children from eastern villages showing the highest malaria parasite and spleen rates. Significant regional differences were also found in mosquito species composition, mosquito abundance, sporozoite rate, and human blood index.

Mosquito abundance was highest in the west and central regions; the sporozoite rate and human blood index were highest in the east, where malaria transmission was also more intense. The results of this survey indicated that soil type and proximity of villages to the River Gambia were not correlated with malaria prevalence levels, although they were highly correlated with vector abundance. It is postulated instead that the length of the transmission season is a more important determinant of malaria transmission (Thomson et al., in press). Both the historical GAC NDVI data sets now available and the 1991, 3-day composite LAC NDVI data set (captured by the Centre Suivie de Ecologique in Dakar, Senegal) can now be used to explore this hypothesis. This information will be incorporated into a model of malaria transmission dynamics for The Gambia the components of which will be identified and analyzed within an appropriate GIS framework.

The ODA's Natural Resources Institute is keen to encourage the wider use of satellite data through their LARST initiatives in The Gambia, which currently receive METEOSAT primary products through a METEOSAT primary data user system. This will permit the production of 10-day cold cloud duration images (a proxy for rainfall). A METEOSAT meteorological data distribution (MDD) receiver system was installed in June 1994, courtesy of the U.K. Meteorological

Office and NRI/ODA. NRI is seeking ways to establish facilities which will enable real-time reception of the NOAA-AVHRR satellite data to assist environmental monitoring and resource management activities in the country (including weather, livestock, tsetse habitat, and factors relating to fisheries resources). While the METEOSAT data has potential value for predicitng changes in malaria prevalence throughout the country, its value would be increased with the availability of the NOAA–AVHRR imagery.

If a model of malaria transmission dynamics can be developed here, based on information from the satellite images, its application in other semi-arid regions where there is little information from ground-based studies may be particularly useful.

How can such tools inform malaria control policy within the primary health care system? Research on the delivery of a decision support facility using RS and GIS must be simultaneously linked to the process of developing the malaria model. Our concordant project aims to review the potential for using a LARST system in malaria monitoring and control for use by the Ministry of Health and non- governmental organizations, in planning the most appropriate malaria control practices.

Using RS/GIS for Assessing the Health Impact of Development Projects

The application of remote sensing images to individual development projects or programs raises further considerations of scale. High resolution images may be useful in mapping the project area and determining potential hazards (Walter 1994). Low resolution images will be of limited value over a small area, but it is important to realise that environmental and health impacts of development programs may not be confined within the project perimeter. They will invariably have some "upstream" and "downstream" effects.

At a macrolevel, rainfall throughout the Nile subbasins and Sudan is of crucial importance to natural resources development along the whole course of the river. Studies on the palaeobotany of Sudan (Wickens 1975), and historical analysis of the changing nature of rainfall in the region, suggest that changes in rainfall in the middle of this century represent 25% of the total variation over the past 20,000 years (Hulme 1989). Such increasing variability and the resulting ecological instability could have very significant effects on the epidemic nature of vector-borne disease. For example, the anomalous rainfall that destroyed the existing *Acacia-Balanites* woodlands in the south can be linked to the current Sudanese epidemic. Such rapid climatic change, however, is unlikely to be causally divorced from the regional development process. Arguments have recently been made that climate should no longer be seen as a fixed boundary condition unaffected by social and economic change, but rather should be viewed as a renewable resource, subject to management criteria, and incorporated into the planning cycle.

A broader perspective view of water shed management must then be taken. However, the need to do so has come at a time when, in the case of Sudan, access to information from ground-based rainfall stations has declined drastically. In the south, data is available from only two sites, Juba and Malakal, compared to over 100 sites in the 1950s. Fortunately, experiments with satellite-derived areal rainfall estimates have been shown to be preferable to rain-gauge data for inputs into hydrological models of a whole catchment (Milford 1989).

Future Earth Observation Satellites

Direct satellite reception is rapidly becoming a powerful, cost-effective, and easily accessible source of data. Moreover, preparations are well advanced for further fleets of satellites, designed to monitor a variety of variables, for a better understanding of the changing global environment. Notable among these are the satellites of the U.S. Earth Observing System (EOS), which will also provide locally accessible data. The quantity of data which is produced by satellites is scheduled to increase by one million times over the next decade. This major resource warrants better scrutiny and utilisation by scientists.

Conclusion

Health sciences researchers have often left considerations of remote sensing and climate modelling to the environmental science lobby. However, quality of health is intrinsic to issues of environmental management and the development process. National boundaries are no barrier to the effects of poor environmental practices. A more integrated assessment approach to development program planning, including health impact, must now be taken. The remote sensing community is keen to increase the range of applications addressed by the new generation of satellites. It is now up to the health community to identify and demand specific types of information from this powerful, yet underutilized, resource.

Acknowledgments

This paper was prepared while the principal authors were engaged by the Liverpool Health Impact Programme. We are grateful to the program manager, Dr Martin Birley, for his support. The Liverpool Health Impact Programme is financed by the Health and Population Division of the Overseas Development

Administration (U.K.). The contributions of our colleagues, R. Ashord, P. Milligan, and M. Service in planning the investigations for the lieshmaniasis and malaria studies is acknowledged. We would also like to thank Dr Pandu Wijeyaratne formerly of the Health Services Division, International Development Research Centre, Canada, for inviting us to participate in this workshop.

References

Ashford, R.W.; Thomson, M.C. 1991. Visceral leishmaniasis in Sudan. A delayed development disaster? Annals of Tropical Medicine and Parasitology,. 8(5), 571–572.

Bonifacio, R.; Dugdale, G.; Milford, J.R. 1993. Sahelian rangeland production in relation to rainfall estimates from Meteosat. *International Journal of Remote Sensing*, 14(14), 2695–2711.

Connor, S.J.; Thomson, M.C. 1994. Towards an understanding of the "Eco-Epidemiology" underlying the current leishmaniasis epidemic in southern Sudan: the role of remote sensing and geographical information systems in decision support. Consultancy Report to MSF Holland. 31 pp.

Connor, S.J.; Allen, P.M. 1994. Policy and decision support in sustainable development planning: a "complex systems methodology" for use in sub-Saharan Africa. Project Appraisal, 9 (2), 95–98.

Dugdale, G.; Milford, J.R. 1986. Rainfall estimates over the Sahel using METEOSAT thermal infrared data. Proceedings of ISLSCP conference on parameterization of land surface characteristics. 1985 ESA SP 248.

Eastman, J.R.; Toldano, J.; Jin, W.; Kyem, P.A.K. (in press). Participatory multi-objective decision making in GIS. Proceedings, AutoCarto XI.

Engelen, G.; Schutzelaars, A.; Uljee, I. 1992. An integrated strategic planning and policy framework for Senegal. Working Paper 925000/0071. Research Institute for Knowledge Systems, Maastricht, The Netherlands.

FAO. 1990. Satellite remote sensing in support of early warning and food information systems, strengthening national early warning and food information systems in Africa, Annex 7, FAO Workshop, Accra, Ghana, 23–26 October 1989.

Hastings, D.A.; Clark, D.M. 1991. GIS in Africa: problems, challenges and opportunities for cooperation. *International Journal of Geographical Information Systems*, 5, 29–39.

Heilkema, J.U.; van Herwaarden, G.J. 1993. Operational use of ARTEMIS data for food security and locust control. FAO Remote Sensing Centre, Rome, Italy.

Holben, B.N. 1986. Characteristics of maximum-value composite images from temporal AVHRR data. *International Journal of Remote Sensing*, 7, 1417.

Howell, P.; Lock, M.; Cobb, S. 1988. The Jonglei Canal — impact and opportunity. Cambridge Studies in Applied Ecology and Resource Management, Cambridge University Press, Cambridge, UK.

Hulme, M. 1990. The changing rainfall resources of Sudan. Transactions of the Institute of British Geographers,. 15, 21–34.

Hutchinson, C.; Toldano, J. (in press). Guidelines for demonstrating geographic information systems based on participatory development. *International Journal of Geographic Information Systems.*

Linthicum, K.J.; Bailey, C.L.; Tucker, C.J.; Mitchell, K.D.; Logan, T.M.; Davies, F.G.; Kamau, C.W.; Thande, P.C.; Wagateh, J.N. 1990. Application of polar-orbiting, meteorological satellite data to detect flooding of Rift Valley Fever virus vector in mosquito habitats in Kenya. *Medical and Veterinary Entomology*, 4, 433–438.

Malingreau, J.P. 1991. Remote sensing for tropical forest monitoring: an overview. In Remote Sensing and Geographical Information Systems for resource management in developing countries, edited by A.S.Belward and C.R. Valenzuela, Dordrecht: Kluwer, 253–278.

Milford, J.R. 1989. Satellite monitoring of the Sahel. *Weather*. 44, 77–81.

Morris, R.S.; Sanson, R.L.; McKenzie, J.S.; Marsh, W.E. 1993. Decision support systems in animal health. Proceedings of the society for Veterinary Epidemiology and preventative medicine (M.V. Thrusfield editor), Exeter, 31 March – 2 April. pp. 188–199.

Perry, B.D.; Kruskas, R.L. 1993. Information management technology to strengthen veterinary services: an appraisal of the systems available and their uses in developing countries. Expert consultation on the need for information systems to strengthen veterinary services in developing countries. Report, FAO, Rome.

Quate, L.W. 1964. Phlebotomus sandflies of the Paloich Area in the Sudan (Diptera, Psychodidae). *Journal of Medical Entomology*, 1, 213–268.

Rajan, M.S. 1991. Remote sensing and geographical information systems for natural resource management. ADB Environment Paper, No 9. Asian Development Bank, Manila, PH.

Rogers, D.; Randolph, S.E. 1991. Mortality rates and population density of tsetse flies correlated with satellite imagery. *Nature,* 351, 739–741.

Schorscher, J. 1991. Entomological survey of the sand fly vector of kala azar in Western Upper Nile, Southern Sudan. Report to MSF-Holland March–July, 34 pp.

Sear, C.B.; Williams, J.B.; Trigg, S.; Wooster, M.; Navarro, P. 1993. Operating satellite receiving stations for direct environmental monitoring in developing countries. In Proceedings of the International Symposium on Operationalization of Remote Sensing. April. ITC, Enschede, Netherlands, 8, 293–302.

Thomson, M.C.; D'Alessandro, U.; Bennett, S.; Connor, S.J.; Langerock, P.; Jawara, M.; Todd, J.E.; Greenwood, B.M. (in press). Malaria prevalence is inversely related to vector density in The Gambia, West Africa. *Transactions of the Royal Society of Tropical Medicine & Hygiene.*

Thomson, M.C.; Connor, S.J.; D'Alessandro, U.; Langerock, P.; Jawara, M.; Aikins, M.; Bennett, S.; Greenwood, B.M. (submitted) Geographical determinants of malaria transmission in The Gambia. *Social Science and Medicine.*

Wadge, G. ed. 1994. Natural hazards and remote sensing. Proceedings of a conference entitled Natural hazard assessment and mitigation: The unique role of remote sensing, 8–9 March 1994, Royal Society, London, UK.

Walter, L. 1994. Natural hazard assessment and mitigation from space: The potential of remote sensing to meet operation requirements. In Natural Hazards and Remote Sensing, G. Wadge, ed., Proceedings of conference held on 8–9 March 1994 at the Royal Society, London, UK, pp. 7–12.

WHO-TDR. 1993. TDR News. Newsletter No 43. WHO Geneva, Switzerland.

Wickens, G. 1975. Changes in the climate and vegetation of the Sudan since 20 000 BP. *Boissera*, 24, 43–65.

Williams, J.B.; Rosenberg, L.J. 1993. Operational reaction, processing and application of satellite data in developing countries: Theory and Practice. In Proceedings of the U.K. Remote Sensing Society, Annual Conference, Towards Operational Applications. Chester, UK.

Wood, B.L.; Beck, L.R.; Washino, R.K.; Hibbard, K.A.; Salute, J.S. 1991. Estimating high mosquito-producing rice fields using spectral and spatial data, International Journal of Remote Sensing,. 13(15), 2813–2826.

Geographical Information Systems for the Study and Control of Malaria

Gustavo Brêtas[1]

Introduction

Vector-borne diseases require an intermediate living agent for their transmission. They are a heavy burden on human populations, a major cause of work-loss, and a serious impediment to economic development and productivity. Their epidemiology is influenced by attributes of their vectors, which in turn are closely linked to environmental conditions.

Over the past decades, the increased demands upon the landscape for food and shelter and an increased number of by-products of man's living environment have led to unparalleled changes. Some of these changes have led to an increase in the distribution of several vector-borne diseases, including malaria.

Malaria is a serious vector-borne disease affecting a greater proportion of the world's population than any other vector-transmitted disease. Over two hundred million cases of malaria occur every year, and the number is increasing. Large areas of regions where malaria had been controlled are now suffering again from this significant public health problem.

The increase in malaria prevalence is determined by several factors: mosquito resistance to insecticides, parasite resistance to drugs, changes in land-use patterns, and reductions in funding and manpower dedicated to control activities. Most of the determinants are heterogeneously distributed, changing over both space and time. Factors such as topography, temperature, rainfall, land-use, population movements, and degree of deforestation have a profound influence on the temporal and spatial distribution of malaria vectors and malaria.

Despite its importance, the study of environmental determinants of malaria has been hampered by the difficulties related to collecting and analyzing environmental data over large areas, and to the speed of change in the malaria epidemiological situation. Geographical Information Systems (GIS), Remote Sensing (RS), and Geographical Positioning Systems (GPS) are important new tools for the study and control of malaria.

[1] Depto. Epidemiologia, Instituto de Medicina Social, Universidade do Esatado de Rio de Janeiro, Brasil

GIS for the Study and Control of Malaria

Over the past 20 years, researchers have been developing automated tools for the efficient storage, analysis, and presentation of geographical data (for example, Aronoff 1989). This rapidly evolving technology has come to be known as "geographical information systems" (GIS). These systems, resulting from the demand for data and information of a spatial nature, may be widely used across varied scientific fields. Their ability to use topological information and spatial analysis functions distinguish them from a number of other information systems.

Geographical information systems may be summarily defined as a constellation of software and hardware tools capable of integrating digital images for the purpose of dealing with geographically-localized data.

Large amounts of information are necessary for almost all aspects of malaria control programs. GIS offer the ability to process quantities of data beyond the capacities of manual systems. Data is stored in a structured digital format, which permits rapid data retrieval and use. In addition, data may be quickly compiled into documents, using techniques such as automatic mapping and direct report printouts (Bernhardsen 1992).

Data Integration

GIS facilitate the integration of quantitative malaria determination and control data with data obtained from maps, satellite images, and aerial photos. Frequently, socioeconomic data and qualitative information on health facilities have a spatial basis, and can also be integrated with malaria data from the same areas. The integration of operational and logistical data for malaria control program planning with epidemiological data will serve to strengthen both the epidemiological analysis and the planning and execution of control programs.

Stratification (Spatial Decision Support)

Malaria stratification aids in the development of community-based malaria control programs, by accumulating past experiences with and solutions to different factors associated with malaria outbreaks (Orlov et al. 1986). Stratification can also point to the existing inequalities in resources, allowing for a more equal and homogeneous distribution of available resources (Kadt and Tasca 1993).

Stratification of malaria emerged as a strategic approach in Latin America in 1979. Since the 1980s, it has become recognized as a useful tool for attaining an objective epidemiological diagnosis, and as a basis for planning strategies of malaria prevention and control.

The ability of GIS to deal with large data sets and to incorporate satellite images increases the feasibility of studying the environmental determinants of malaria. The epidemiological mapping of high-risk areas of malaria transmission has helped national level malaria control programs to recognise those populations and geographical areas where it is possible to identify the main determinants of malaria morbidity and mortality.

One of the main functions of GIS is the selection of areas according to some specified characteristics, in a search for regularities. These characteristics may be both or either spatial or nonspatial data from a map or database within the GIS. Data may be selected according to geographic criteria or content. Most GIS incorporate the use of Structured Query Language, which allows logic and arithmetic operations. Queries may be built to stratify areas according to inundation potential, proximity to treatment sites, availability of transport, incidence of malaria, quality of housing, and so on. Characteristics may also be combined: one could ask the GIS to find areas where the houses are of poor quality and the drainage of the terrain slow, as possible candidates for a control strategy with emphasis on drainage instead of spraying.

The stratification of areas according to their epidemiological situation is also useful to development agencies implementing agricultural projects. Many such projects require environmental and health impact assessments, which could be used as a working base by malaria control programs in the early stages of land occupation. At the same time, the projects would benefit from the control programs' previous data. It is thought that a GIS would be an adequate tool to integrate these different data-sets and information.

Heuristic Modelling

Maps are heuristic models, used to communicate, interpret, and explain data. They aid in the visualisation of differences, clustering, heterogeneity, or homogeneity within data. Spatial patterns can be perceived and correlations visualized through the use of maps. Symbols and colours can communicate detail or the relative importance of certain features.

Malaria control program staff tend to be familiar with maps, using them for their daily activities. Maps can consequently be used to communicate ideas and explanations about the determinants of malaria and strategies of control. A map's heuristic potential can also be exploited as the media of communication between a control program and the public.

Analysis

A number of analysis methods may be used on data acquired through a GIS. Data may be selected according to geographic criteria or content using some form of Structured Query Language. Geometric operations involve the computation of distance, areas, volumes, and directions. Flow analyses calculate geometric computations to find the shortest routes between designated points. This type of analysis is frequently important for the planning, logistics, and operation of malaria control programs.

Statistical operations are performed primarily on attribute data, but may also be applied to some types of geometric data. Most GIS support a range of statistical operations, including sum, maxima, minima, average, frequency distribution, bi-directional comparison, standard deviation, and multivariate analysis. Pattern recognition and statistical modelling are also incorporated in some GIS (Bernhardsen 1992), allowing for the identification of spatial and/or temporal clusters of malaria occurrence. The relationship between these clusters and environmental or socio-economic characteristics can then be investigated.

GIS also open new possibilities of data analysis not limited to spatial analysis. Data generated within GIS, distances, areas, and selections based on spatial criteria, can all be used as inputs to statistical modelling.

Global Changes and Malaria

It is expected that changes in the global environment will lead to changes in malaria occurrence patterns. These changes are currently being studied and modelled, and their impact on malaria can now be examined. For example, researchers are looking at the impact of increased temperatures on the spatial distribution of malaria, and on the impact of deforestation and temperature increases on the latitudinal limits and intensity of highland malaria.

The occurrence of malaria under different land occupation strategies can be studied with the combination of remote sensing and GIS. The results could be used as guidelines for new development projects in areas receptive to malaria.

Integration with other Information Systems

WHO technical report no. 839 (1993) recognizes that

> a change from specialised information systems for malaria towards integrated health information systems is essential if the general health services are to take full responsibility for managing malaria. An integrated health information system will allow the malaria problem to be related to other health problems...

Taken further, it can be seen that the main determinants of malaria are not restricted to health. Consequently, there is a need to integrate other types of data into a malaria information system. Since data is collected to fulfil different objectives and with different formats, one of the few possible strategies for its integration is the use of a geographic base. A strategy for the comprehensive integration of data for the planning and execution of health promotion activities (access to water, sanitation, housing, and so on) has been developed using a GIS as a central tool (Project Blade Runner, see Kadt and Tasca 1993). A similar instrument could and should be developed by malaria control programs.

Tools Complementary to GIS

Satellite Remote Sensing (RS)

Satellite imagery in digital format allows for the acquisition of environmental data and land occupation patterns and features over large areas. Sensors in satellites record multispectral data from different wave bands in digital format. Different features of the terrain reflect differently in each waveband, allowing for their recognition in the images. The digital image is fed into the computer, where it is stored. The digital images can then be displayed and further processed to extract the desired information. They may also be integrated with other types of data and information within a GIS.

The information found within a digital image is contained in a grid constructed of spatial units called pixels. Each pixel number is related to colour intensity and brightness. The main limitations of satellite images are cloud cover[2] and resolution. Even with the best resolutions available (pixels < 30m), it is not possible to see houses, to adequately classify some types of agricultural practice, or to localise some breeding sites. Some of these problems may be circumvented using satellite navigation receivers.

By using the data from the different wave bands, it is possible to identify and track environmental characteristics and changes useful to the study of malaria and other tropical diseases. Vegetation, land use patterns, surface waters, quality and humidity of the soil, roads, built-up areas, and climatic changes may all be monitored by satellite.

Satellite Navigation System (SNS)

The satellite navigation system (SNS) was originally designed to enable a user to obtain an instantaneous three-dimensional position, anywhere on the earth, at any time, under any weather condition. Clock information in satellites is emitted by radio wave.

The difference between the time when a message is sent and the time when the message is received allows for the calculation of the distance between the

---

[2]Earth Resources Satellites pass over the tropical areas early in the morning, a period when clouds and mist are most prevalent.

receiver and the satellites. Since the orbits of the satellites are known, exact positions can easily be calculated.

The SNS can be used in a number of ways to calculate absolute and relative positions with varying degrees of accuracy. The complete system consists of 24 satellites which are distributed in such a way that an adequate number of satellites are available for positioning at any time.

When associated with a GIS, a SNS receiver is a powerful mapping tool. It can provide quick and accurate positioning of terrain features and dynamic mapping. The data received can be transferred to a computer and read by a GIS, where it is transformed into map format.

Project InfoMal[3]

Objectives

This project, funded by the International Development Research Centre (IDRC), has been designed to assess the environmental determinants of malaria in the Brazilian Amazon using satellite remote sensing, geographical information systems, and a satellite navigation system. Its objectives are:

- To compare the risks of malaria infection under different environmental situations, controlling for the impact of socioeconomic determinants,
- To compare the risks of malaria infection within different socioeconomic groups, controlling for the impact of environmental determinants,
- To derive the rank of risk situations from the relationship of environmental and socioeconomic factors, and
- To build a rank of malaria risk situations for agricultural areas in the Amazon. It is also intended to organise an information system based on a regional GIS.

Methods

The methodological approach relies upon the integrative capabilities of GIS (Aronoff 1989). The data will be submitted to further multivariate statistical analysis to seek the best predictive indicators of malaria risk.

Different data sets, environmental data, and land occupation patterns from satellite imagery, coordinates of houses, and features of the terrain obtained with

---

[3]More information about the project can be obtained by sending an e-mail to gbretas@lshtm.ac.uk or writing to Gustavo Bretas LSHTM/THEU, Keppel St., WC1E 7HT, London, UK, fax 071-9272216.

SNS, map data from military and road maps, data from the malaria control program, and collected socioeconomic, household microenvironment and malaria data will be integrated into the GIS. The analysis will be conducted in parallel within the GIS and with statistical software. Results from the preliminary analysis will feed back to further analysis (for example, distance from water bodies, amount of deforestation, and land use will be calculated within the GIS and used in the multivariate modelling of malaria determination).

The Information System

In the past, information systems for malaria control were designed for use by eradication programs. The data was analyzed and used centrally with little or no feedback to the source of the information and with no impact on local activities.

The World Health Organization soon recognized, however, that information systems are a vital element for strengthening the national and local capacities for assessing malaria situations, for selecting appropriate control measures, and for adapting activities to changes in the malaria situation. It was seen that information systems should, therefore, be reoriented to deal with malaria, and decentralized in such a way that information is available to and used by those who need it.

Since malaria occurrence is influenced by numerous phenomena outside the habitual framework of health systems (including population movements, environmental changes, and agricultural practices), the information system should be able to incorporate and use these different types of information.

Each of these different types of data and information has a spatial basis. It was, therefore, decided that it was important that an information system to be used for malaria control should be able to deal with geographical aspects of information.

The Necessary Capabilities

It was decided that the system chosen should be able to:

- Use to the existing databases of the malaria control program,
- Be open to the future incorporation of data from other sources (rs, agricultural, roads, and so on),
- Deal with the spatial aspects of this information,
- Produce routine epidemiological evaluations,
- Stratify areas based on a set of rules,
- Automate maps,
- Produce feedback to the peripheral level, and
- Produce feedback to the general public.

Main Constraints of and Problems With the Current Use of GIS

Cost

Costs are currently the main constraint to the use of GIS. Software is relatively expensive in relation to the budgets of malaria control programs. Expansion of GIS use will probably decrease its cost. In the meantime, a compromise would be to use available shareware, or to have a software tailored for use in malaria control programs by a nonprofit organization.

Adequate Training

There exists at present a lack of trained personal. The new windowed interfaces, however, are easier to use and will speed up the process of training. Computing skills are useful in the labour market; consequently, staff from control programs should be willing to be trained.

GIGO (Garbage In, Garbage Out)

GIS is not a tool designed to increase the quality of data. Frequently, much of the data collected by malaria control program staff are not used. GIS use could lead to a relaxation in data collection and consolidation. It is necessary to review all the steps in the information flow to guarantee quality and adequacy.

Misinformation and Misinterpretation

The powerful tools of GIS can easily lead to misinformation and misinterpretation, particularly by someone unfamiliar with their use. Ecological fallacies, problems of scales, and propagation of error are frequent, and should be given serious consideration (Monmonier 1991). The same problem is also true with other types of software which have helped to advance our capacity to study and control malaria, such as spreadsheets, graphic software, and data analysis software. GIS is not a magical solution to all the information difficulties of malaria control, but is a powerful tool capable of transforming the way with which information is dealt.

References

Aronoff, S. 1989. Geographical information systems. Management perspective. WDL Publications Canada, 1989.

Bernhardsen, T. 1992. Geographical information systems. Viakt IT, Arendal, Norway.

Kadt, E.; Tasca, R. 1993. Promovendo a Equidade: Um novo enfoque com base no setor saude. Editora Hucitec e Cooperacao Italiana em Saude, Sao Paulo-Salvador, BR.

Monmonier, M.S. 1991. How to Lie with Maps. Chicago. University of Chicago Press, IL, USA.

Orlov, V.S.; et al. 1986. The concept of stratification of territories and its practical implications. WHO, Geneva, Switzerland. WHO/MAL/86.1032.

World Health Organization. 1993. WHO study group on the implementation of the global plan of action for malaria control. WHO, Geneva, Switzerland. WHO Technical Report Series, No. 839.

Spatial Analysis of Malaria Risk in an Endemic Region of Sri Lanka

D.M. Gunawardena,[1a] Lal Muthuwattac,[b] S.Weerasingha,[a] J. Rajakaruna,[a] Wasantha Udaya Kumara,[a] Tilak Senanayaka,[c] P. Kumar Kotta,[c] A.R. Wickremasinghe,[d] Richard Carter,[e] and Kamini N. Mendis[a]

Introduction

Malaria is a major public health problem in Sri Lanka, with almost 300,000 infections being reported yearly within a population of 16 million. As much as two-thirds of the entire national public health budget is spent on controlling malaria, and the disease constitutes the fourth highest cause of hospital admission in the country. Malaria has made a major impact on the health, economy, education, and general development of the population (Administration Report 1991). Although in the past *P. falciparum* malaria was a rarity, its vector is now almost as prevalent as that of *P.vivax*; further, with chloroquine-resistant strains spreading rapidly, malaria promises to become a more serious problem in the future.

Malaria has always been endemic through much of the country, but especially in the dry zone. The central hill country has remained free of the disease. The malarious areas face perennial transmission with two seasonal rises related to the rainfall pattern: June–July and December–January, associated with the southwestern and the northeastern monsoonal rains, respectively. Lately, a significant increase in malarial morbidity has been reported, probably due largely to the increase in *P.falciparum* infections, many of which are also chloroquine resistant.

[1a]University of Sri Lanka, Nawala, Sri Lanka.

[b]Division of Mathematics and Management of Technology, Open University of Sri Lanka, Nawala, Sri Lanka.

[c]South Asian Cooperative Environmental Program, Colombo, Sri Lanka.

[d]Department of Community Medicine, Faculty of Medicine, University of Peradeniya, Sri Lanka.

[e]Division of Biological Sciences, ICAPB, University of Edinburgh, Scotland, UK.

A combination of factors contribute to the difficulties associated with trying to achieve a reduction in malaria transmission. These include:

- Environmental and geographic features of the area, such as climate, land use patterns, and development of irrigation schemes;
- The movement of people and the creation of new settlements;
- Behavioral aspects relating to the population;
- The health care delivery system; and
- The socioeconomic and educational status of the population (Ministry of Health 1987–1991).

Even in moderately endemic areas of the country, it is well known that malaria infections are not homogeneously distributed; some individuals and families are subject to repeated malaria infections, while others experience no infections at all (Gamage–Mendis et al. 1991). This clustering of malaria infections seen within a population led to an investigation of malaria risk factors in endemic communities in Sri Lanka. One such risk factor was house construction type; poorly built houses with mud walls and thatched roofs afforded a significantly higher risk to their inhabitants than houses constructed of plastered brick walls and tiled roofs (Gunawardena et al. 1994.).

In the present study, human malaria infections occurring in an endemic population in Kataragama in southern Sri Lanka were monitored; particular attention was paid to the spatial and geographical features of the area. This enabled an analysis of the malaria risk of this community with respect to three potential risk factors: location of houses in relation to distance from the forest edge and a source of water, and the construction type of houses.

Materials and Methods

Background

This study is part of a research project on the microepidemiology and clinical aspects of malaria, which was conducted in Kataragama, a highly endemic region for both *P.falciparum* and *P.vivax* in southern Sri Lanka. The study was carried out over an 18-month period from mid-January 1992 to July 1993. The total population of this study consisted of 1875 residents in 423 houses, in a cluster of eight contiguous villages. The area was geographically and demographically defined by reconnaissance and census.

The Setting

The study site is situated in the country's dry zone, where most people depend on "Chena" (shifting-bush-fallow system) cultivation as a traditional agricultural practice. Most of the farmers live close to the forest edge and/or a source of water. Of the 8 villages, one is a small town situated in the middle of the study area. The other villages are surrounded by the forest. A river (the Menik ganga) flows through these villages. The forest is typical of dry zone forests with dry mixed evergreen vegetation. The forest edge consists mainly of thorny bushes, extending gradually to larger trees (scrub to dense forest). The area is characterized by a rainy season from October to February, and a dry season from June to September. In the dry season the river dries up, creating rock pools on the river bed.

Mapping

Aerial photographs of the study area taken in 1993 at an altitude of 10,000 meters were obtained from the Survey Department. These photographs were enlarged four times to obtain a map having a scale of 1:5000 with real world coordinates. The exact location of the houses, roads, land use, and forest cover, and significant water bodies like rivers, small streams, and reservoirs, were marked on the map. The accuracy of the exact location of such landmarks was confirmed by geographic reconnaissance (GR) carried out during the preparatory stage of the study. The maps were digitized and the GIS package ARCINFO was used to obtain the nearest distance of a particular house from the forest edge and a source of water. In the case of flowing water bodies, the centre of the flow was used to measure the distance from the water source to the house. All the houses in the study area were numbered, and every house and individual residents in a house were assigned a unique identification number to enable us to monitor malaria infections.

Case Detection

Monitoring of all malaria infections in the study area was carried out by two methods. Passive case detection (PCD) entailed the presentation of fever patients for blood film examination and treatment, either at the Field Research Station which is very close to the study area, or at the Government Hospital, Kataragama, which is about 3km from the study area. Active case detection (ACD) was done by a series of six mass blood surveys at intervals of 2–3 months. ACD was carried out using three to four teams, with each team doing a house-to-house survey and obtaining thin and thick blood films from all residents. A diagnosis of malaria was established by demonstration of malaria parasites on microscopic examination of thin and/or thick blood films stained with Giemsa stain. It is estimated that a very high proportion of all malaria infections occurring in the population were recorded using these methods.

House Types and Malaria Incidence Rates

All houses in the study area were categorized into two broad classes according to the type of house construction. This categorization was based on the nature of, and building material used for, the walls and roof. Those houses which had incomplete and/or mud walls and roofs made of thatched coconut palms were classified as being of poor construction type, and those with complete plastered brick walls and tiled or corrugated iron roofs were classified as being of good construction type.

The nearest distance from a source of water and from the forest edge to each house was determined using the GIS software package ARCINFO. In estimating the incidence rates of malaria for each house, those houses with two or fewer residents were excluded from the analysis as they may have given unstable rates, thereby leading to errors in inference.

Definitions

The incidence rate in a house was defined as the total number of malaria infections which occurred in individuals resident in the house during the 18-month period, divided by the number of residents in the house.

The average incidence rate for a group of houses (e.g., of a particular construction type) was estimated as the mean incidence rate of the house in that group.

The incidence rate of a population was defined as the number of infections which occurred in that population divided by the size of the population.

Results

Of a total of 423 houses and 1,875 residents in this population, only 343 houses and their resident population of 1,744 were considered for analysis, the rest being houses with fewer than three residents.

During the 18-month period, 1,579 malaria infections were detected by microscopic examination of blood films in the 343 households; 913 infections were due to *P.vivax* and 666 to *P.falciparum*. The incidence rate during the 18 months (number of infections per person) in the selected population was 0.91. Of the total of 343 houses, 182 were of poor construction type and 161 of good type. The distribution of the two types of houses within the area was not uniform, the houses of better construction type tending to cluster around principal roads. Most of the poorly built houses were scattered across the entire area. Residents of this area were almost equally distributed between the good (838) and poor (906) house types.

A significant difference was found in average malaria incidence rates among the populations resident in good and poorly constructed house types, being 0.51 and 1.23 infections per person, respectively. The risk was thus 2.5-fold higher in inhabitants living in the poorly constructed houses (95% C.I. – 2.19, 2.93).

There was a significant negative correlation between the incidence rates of all houses and the distance of the houses from both the forest edge and a source of water (r= –0.25; p=0.0001 and r= –0.12; p=0.0264, respectively). The location of the houses of the two different construction types were not significantly different with respect to their distance from water (p=0.7251), but they were with respect to distance from the forest edge; the poorly constructed houses were significantly closer to the forest than were the good (p=0.001).

Since house construction type was found to be a determinant of malaria risk and the distribution of the two types of houses were spatially clustered, malaria risk was analyzed in relation to the distance from the forest edge and a source of water separately for the two house types. In neither house type were the incidence rates correlated with the distance from the forest edge (r= –0.09; p=0.2292 for houses of poor construction type; r= –0.12; p=0.1431 for houses of good construction type). This implies that the association between malaria risk and the distance of houses from the forest edge was confounded by the type of house construction.

With respect to the malaria risk in relation to the distance from water, a significant correlation was obtained for poorly constructed houses, the risk being greater in houses closer to a source of water (r= –0.31; p=0.0001). In houses of good construction type, however, a marginally significant correlation was found with distance from the water, but in the reverse direction (r=0.14; p=0.0676).

Discussion

Malaria has been prevalent in most parts of the tropical and subtropical world, including parts of Africa, Asia, and Central and South America, for a very long period of time. The disease has been the single most important cause of morbidity in some of these areas. Yet, a desirable level of control has not been achieved during the past two decades. Prospects appear bleak for the near and distant future as very limited options for control are currently available. Resources are also becoming scarce. Disease control in the future is likely, therefore, to benefit from concerted efforts, including those based on an understanding of the microepidemiology of the disease in a given situation. This finding motivated this research team to conduct studies on the microepidemiology and clinical aspects of malaria, paying particular attention to ecological factors.

In a previous study, it was demonstrated that malaria infections in an endemic area were clustered in particular individuals and households (Gamage–Menids et al. 1991). It has also been shown that the incidence of malaria is associated with the construction type of houses. Residents of poorly built houses with incomplete mud walls and thatched roofs were at a higher risk of acquiring malaria than those living in better built houses with complete plastered brick walls and tiled or corrugated iron roofs (Gunawardena 1994).

The present study supports earlier findings of an association between malaria risk and house construction type, this study taking place in a different population and conducted over a different period of time. In addition, this study presents a more detailed examination of two other ecological factors which could determine the malaria risk, these being the proximity of the house to the forest edge and a source of water. Both these entities could, by their association with the mosquito vector, be expected to influence the malaria risk in an endemic community. The forest edge is thought to provide ideal resting places for the mosquito, thus influencing its survival and the level at which malaria transmission is sustained. The water bodies and rivers examined in this analysis are known to be potential breeding places of the mosquito vector. Thus, proximity to both might be expected to impose an added risk for acquiring malaria.

The findings of this study indicate that the risk of contracting malaria was on average 2.5-fold greater for a resident of a poorly constructed house than for one who lives in a well built house. In the first study which demonstrated an association between house types and malaria risk (Ministry of Health 1991), higher densities of mosquitoes were measured and found resting within poorly constructed houses, than in houses of better construction type.

This finding suggests that the risk of being bitten by a mosquito might have been greater within the poorly constructed houses, given that the construction type of the house itself afforded a better respite for the vector. It is nevertheless believed that several other factors associated with the type of house, such as socioeconomic status, education levels, and occupations of the residents, might independently influence the malaria risk. Thus, for the purpose of the discussion that follows, the house construction type is considered as much a marker of the general standard of living, as of the construction type itself.

Overall, there was a significant negative correlation between the incidence of malaria and the distance of the house from the forest edge and a source of water, when all houses in the community were considered. This was not evident when the analysis was performed separately for each of the two house construction types. In neither house type was the malaria risk correlated significantly with the distance from the forest edge. Houses of good construction type were found to be clustered around the major roads in the area, whereas the houses of poor construction were generally more scattered. Thus, most of the houses that were close to the forest edge were of the poor construction type; this is easily explained by the fact that better houses belong to the more affluent, who have the ability to acquire land in relatively more urban parts of the area. It follows, therefore, that because the better built houses were located away from the forest edge, house construction type either by itself or by the factors that are associated with having a poor house, was confounding the apparent relationship between malaria incidence and the location of the house in relation to the forest edge.

Houses of both construction types were distributed fairly equally in relation to the sources of water. The relationship between malaria risk and proximity to a source of water was different for houses of the two construction types. For the poorly constructed houses, a significant negative correlation was found with the distance from water, demonstrating a decrease in malaria risk with increasing distance from the water. This was not so in the better constructed houses, in which such a correlation was not apparent. In fact, a trend in the reverse direction was evident, i.e., a tendency for the malaria risk to decrease as the house moved closer to the water source, although at a borderline level of significance. The water bodies considered here are potential breeding sites for the vector mosquito. The increasing risk of malaria in houses located close to these water sources is therefore to be expected, for the reason that a higher density of mosquitoes must prevail in the vicinity and the probability exists for a higher man-mosquito contact closer to the water.

This appears to be so in the case of poorly constructed houses, but not in houses of the better construction type. It thus appears that the risk imposed by being close to a source of water is transcended by an improvement in the house type, be it because of the construction type itself or the associated life styles or characteristics of its residents. If the trend of a decreasing malaria risk in residents of good house type closer to water reaches a level of statistical significance, it would suggest that other factors pertaining to its residents and the house which have not been examined here are more important determinants of malaria risk than the distance from water.

The association between malaria risk and the location of the house in relation to a source of water has implications for irrigation schemes and agriculture-based developmental projects, both of which involve manipulation of waterways in which malaria vectors breed. They also result in increasing the number of human settlements located close to water. The finding, however, that the higher risk of malaria imposed by being close to a source of water can be overcome by a better house type and/or higher standard of living is encouraging for such projects, which are expected to eventually increase incomes and improve living standards. It also implies that the initial health hazards for settlers in such schemes could be significantly offset by investing in better housing and a generally higher standard of living.

Relevance of GIS to this Study

This study constitutes an example of the use for GIS in health research. The specific advantages offered for this analysis by GIS were as follows:

- The display of the data pertaining to the study area provided an overview of the malaria incidence in relation to geographically and ecologically important entities. It also provided clues for further analysis, e.g., the fact that the better built houses were seen to be spatially clustered around the major roads suggested that house construction type may have confounded the relationship between malaria risk and the two ecological entities — the distances from the forest edge and water; this prompted an analysis of the correlations with these two ecological risk factors separately in the two different housetypes. The program offered the obvious advantage of having reproducibility of high-quality maps for presentation and publication.
- It was possible to obtain accurate distances between selected landmarks which were essential to this study.

The present investigation did not make use of topological features and, therefore, utilized only a fraction of the full potential of a GIS. However, as malaria transmission is dependent on many ecological factors, many more uses for GIS are envisaged, especially in developing models for malaria control.

It is obvious that the usefulness of GIS and the value of inferences made from spatial analyses will depend on the quality of the data used. This analysis was based on data of a very high quality both in terms of accuracy and detail, obtained from a research project. It is probably unrealistic to expect data of such high quality from disease control programs. This study and analysis, therefore, exemplifies the use of GIS primarily as a tool for health research. Our ultimate goal, however, includes the development of a GIS that will be useful as a decision support system for a malaria control program.

In developing a GIS application which effectively can be used in a disease control program, it is essential that the important variables are identified and accurate data obtained. In this regard, it is pertinent to address the following questions:

- Should GIS applications for disease control be scientifically evaluated?
- If these applications are to be evaluated, what criteria should be used?
- At what time point should the evaluation be carried out?

It is apparent that GIS are being used for health by a number of researchers in varied areas of specialization. Their potential has been clearly demonstrated in a number of research projects. However, we feel that if a GIS is to be introduced as a tool to aid control programs, then such applications should be subject to rigorous evaluation. This is essential especially for third world countries such as Sri Lanka, which must make judicious choices given limited resources. For instance, if a GIS is to be used in the malaria control program in Sri Lanka, then computerization of malaria control activities needs to take place.

Such an activity requires an enormous amount of resources, both in terms of hardware and manpower. Unless there is sufficient scientific evidence to show that GIS applications help in reducing malaria transmission via a better decision support system and implementation of such decisions, it would be an arduous task to convince both policymakers and control personnel of the need to utilize GIS application in disease control programs.

The main objective of disease control is to reduce transmission and incidence of disease. Therefore, evaluation of a tool that aids disease control should use the occurrence of disease as the marker. A useful tool should objectively show a reasonable level of reduction of the incidence of disease. What constitutes a reasonable reduction would depend on the program and the perceived threat of the disease.

The application of GIS for health is still in its infancy. Therefore, early evaluations may over- or underestimate the usefulness of GIS as a tool in disease control. However, a mechanism for evaluation of GIS applications should be incorporated in planned stages in the development of such applications in the future. Judicious evaluations would provide useful information that would eventually lead to improvement of the application, increasing the prospects of it being a more useful tool for disease control.

References

Administration Report. 1991. Anti Malaria Campaign, Ministry of Health, LK.

Gamage-Mendis, A.C.; Carter, R.; Mendis, C.; De Zoysa, A.P.K.; Herath, P.R.J.; Mendis, K.N. 1991. Malaria infections are cloistered within an endemic population: Risk of malaria associated with house construction type. *Am. J. Trop. Med. Hyg.*, 45, 77–85.

Gunawardena, D.M.; Mutuwatta, L.P.; Wickramasingha, A.R.; Weerasingha, S.; Rajakaruna, J.; Carter, R.; Mendis, K. N. 1994. Malaria risk factors in an endemic area of southern Sri Lanka. Sri Lanka Association for the Advancement of Science, LK. (Submitted for publication)

Ministry of Health. 1987–1991. Plan of operation for malaria control in Sri Lanka. Anti-Malaria Campaign, Sri Lanka. LK.

Diagnostic Features of Malaria Transmission in Nadiad Using Remote Sensing and GIS

M.S. Malhotra and Aruna Srivastava[1]

Introduction

Nadiad taluka of Kheda district, Gujarat, is situated in western India. Epidemiological investigations demonstrate that malaria is endemic in the state. Kheda district showed the highest incidence of malaria in Gujarat state, with Nadiad taluka consistently being the most affected area. Occasional deaths have occurred due to falciparum malaria, although the area is constantly sprayed with residual insecticides. In 1981, a serious epidemic of *P. falciparum* malaria resulted in the deaths of 31 people in one village alone (Sharma et al. 1986).

An. culicifacies is the main vector of rural malaria. It breeds in stagnant water in fallow fields, rice fields, irrigation channels, field channels, and seepage water collection sites. During the monsoon season, high humidity favours malaria transmission, as the longevity of *An. culicifacies* increases.

A feasibility study using a geographical information system (GIS) was initiated in April 1993 to identify the specific factors responsible for high receptivity and vulnerability to malaria in Nadiad taluka. The aims of this study were to:

- Map the various geomorphological, physical, and climatic characteristics of the region;
- Develop a georeferenced database related to malaria; and
- Identify relevant factors having a direct or indirect bearing on malaria transmission in Nadiad taluka.

Materials and Methods

The economy of Nadiad taluka has been based predominantly on agriculture since 1966 (Chowdhary et al. 1983). The average annual precipitation is about 800mm, all of which falls during a rainy season of 35 days (Ashok and Nathan 1983).

Since stagnation of water for long periods of time is likely to create mosquitogenic conditions, the following factors related to water stagnation were

[1]Malaria Research Centre, Delhi, India.

taken into consideration: soil type, water table, irrigation, surface water bodies, topography, climate, drainage, and hydrogeomorphology.

Topological sheets at a scale of 1:250,000, showing topography and terrain of Nadiad, were obtained from the Survey of India, Dehradun. These sheets were incorporated with satellite imageries, and a composite map of Nadiad was drawn showing important physical features of the region, such as slope, forest/vegetation, rivers, streams, reservoirs, other water bodies, human settlements, roads, and so on. This map was used to extract the relevant features and specific characters of the region.

Findings

Soil Types

Data and maps on types of soil existing in Nadiad were obtained from the appropriate authorities. Soil zones were marked on the maps, which were then digitized. Four main types of soil are present throughout Nadiad taluka.

Table 1. Soil Types and Characteristics in Nadiad Taluka

Soil Type	Number of Villages	Water Retention Capacity
Ankhi Series Drained, noncalcareous, fine loamy and mixed	38	Medium
Ratanpur Series Noncalcareous, fine clay, mixed	52	High
Matar Series Well drained, slightly saline, alkaline type, sandy loam	0	Low
Chaklasi Series Well drained, non-calcareous, course, loamy and mixed	11	Low

In southwestern Nadiad, Ankhi and Ratanpur series of soil were encountered. These soils are comprised of loam and fine clay and are non-calcareous. As this area of Nadiad is located within an extensive irrigation zone that has a network of canals and drains, water stagnation is common. A small region in southern Nadiad is characterized by the Matar soil series. The Chaklasi soil series is also found in small area in the south, as well as in northern Nadiad.

Water Table

Subsoil water is an important factor for the creation of stagnant water pools. Shallow ground water can result in stagnant, humid marshy areas, which encourage mosquito breeding and survival.

Subsoil water has become shallower in several areas of the Nadiad region through the development of irrigation systems. In 1958, the year in which irrigation canals were first introduced into the area, no water tables existed above 1.5m from the surface during premonsoon. With the installation of the irrigation systems, the water table began rising, and by 1981, 475ha of land had water tables located at less than 1.5m from the surface. Similarly, the area with water tables located from 1.5 to 3.0m below the surface increased from 2,643ha in 1958, to 9,360ha in 1981. Other water table depths are shown in Table 2.

Table 2. Rise of Water Table in Irrigated Area

Depth of water table from ground surface (m)	Area Covered			
	ha	1958 % of total	ha	1981 % of total
< 1.5	-	-	475	0.2
1.5 – 3.0	2643	0.9	9360	3.2
3.0 – 6.0	18058	6.1	83338	28.3
6.0 – 9.0	15277	5.2	33500	11.4
> 9.0	257882	87.8	167207	56.9
Total	293860	100	293880	100

Topography

The Nadiad region is characterized by a generally flat topography, restricted natural drainage soils, and a semi-arid climate. The area slopes north–east to south–west, with an average gradient of about 1–1,600m towards the Gulf of Cambay. There are, however, a few isolated local high spots and ridges. Without the existence of an efficient water drainage systems, each of the factors mentioned above can lead to the creation of stagnant water pools, which, in turn, encourage the breeding and survival of mosquitos.

Hydrogeomorphology

Two categories of geomorphology may be found in Nadiad. Flood plains are found near the shores of the Shedi river and its tributaries. This land feature has developed relatively recently, and its moisture content is very high. Alluvial plains exist throughout the rest of the Nadiad region. The moisture content of the Alluvial plains is relatively lower than that found in the flood plains.

Irrigation, Drainage, and Surface Water Bodies

The Mahi river is one of the major river systems in western India. In 1978, a reservoir was constructed on this river for irrigation. The command area of this irrigation system includes the southern half of Nadiad taluka. The abundance of water in irrigated areas is due to seepage, silting, and stagnation, creating innumerable breeding sites for mosquitoes. A thematic map showing details of the irrigation system was prepared and digitized.

Tank irrigation was one of the earliest systems of irrigation to be used in the Nadiad region. Irrigation tanks and ponds, when used together with canal irrigation, act as storage reservoirs. Excess irrigation water is diverted to these tanks, for use for when canal sources become scarce.

Every village in the Nadiad region has at least one pond from which water is gathered. These ponds receive both rain and canal water. In an attempt to fight famine in the Nadiad area, the village ponds were deepened. While this has created a year-round supply of pond water, it has also created a high local humidity level conducive to heavy vector breeding. The information on pond location was extracted from topological sheets and remote sensing composite maps.

Climate

Climatic features such as rainfall, humidity, and temperature have a direct influence on the propagation of mosquitoes and their survival. These data were plotted to study the relationship between climate and malaria transmission.

Results and Conclusions

ARC/INFO was used to overlay eight maps, producing a composite map of soil type, water tables, topology, hydrogeomorphology, irrigation, drainage and surface water bodies, and climate. The resulting georeferenced data analysis divided Nadiad into four water-holding areas.

When this map was superimposed onto a map showing high, low, and average API (number of malaria cases per 1,000 population per year) of 5 years at the local level, however, the resulting picture is difficult to explain: high API villages do not show any visible correlation to the water holding capacity of the four areas of the composite map.

This result demonstrates that to establish determinants of malaria in Nadiad, it is necessary to analyze other data in addition to that generated by a study of water holding capacity. Additional parameters that could be examined include population movement, labour settlement, intervillage migration, rice cultivation, creation of borrow-pits that supports the breeding of *An. culicifacies*, the outreach of health services, and so on.

It was established that malaria was more prevalent in areas with high water tables, the presence of surface water, and canal irrigation systems, than in riverine villages. The former areas are more conducive to the breeding of *An. culicifacies.*

Analysis of climatic data has demonstrated that during years in which the relative humidity (RH) was above 60%, malaria transmission was high, while during years when the RH was below 60%, transmission rates were low. Humidity is, therefore, seen as an important conditioning factor for vector presence, as it governs longevity and, therefore, the reproduction rate of *An. culicifacies.*

It may be noted that the Kheda district in Gujarat does not have a regular malaria transmission pattern. Ground water stagnation is more pronounced where the water table is shallow. In years with favourable climatic conditions, such as high rainfall levels and high humidity, malaria transmission increases within the nonimmune population, creating epidemic situations.

Because of population movement, migration, and poor surveillance, it is difficult to analyze the data at a micro or village level. GIS should, therefore, be applied to macrolevel analysis to:

- Bring out broadly the conditions conducive for vector breeding and malaria transmission, and
- Suggest integrated control methods that are both cost effective and sustainable.

References

Ashok Raj, P.C.; Nathan, K.K. 1983. In: A case study of Mahi right bank canal command area. Gujarat IARI. New Delhi, IN. pp. 19–46.

Chowdhary, R.K.; Michael, A.M.; Sarma, P.B.S. 1983. In: A case study of Mahi right bank canal command area. Gujarat IARI. New Delhi, IN. pp. 1–18.

Sharma, V.P.; Sharma, R.C.; Gautam, A.S. 1986. Bioenvironmental vector control of malaria in Nadiad, Kheda district, Gujarat. *Indian Journal of Malariology,* 4, 23(2), 95–117.

Monitoring Zoonotic Cutaneous Leishmaniasis with GIS

L. Mbarki[1a], A. Ben Salah[a], S. Chlif[a], M.K. Chahed[d], A. Balma[a], N. Chemam[b], A. Garraoui[c], and R. Ben-Ismail[a]

Introduction

Zoonotic cutaneous leishmaniasis (ZCL), caused by *Leishmania major*, poses an important public health problem in Tunisia, with more than 50,000 human cases being recorded since the beginning of 1983 (Ben-Ammar et al. 1984; Ben-Ismail and Ben-Rachid 1989; Ben-Ismail et al. 1987). The disease is transmitted by a vector of the species *Phlebotomus papatasi*; the reservoir hosts are rodents, namely *Psammomys obesus*, *Meriones shawi* and *M. Libycus* (Ben-Ismail et al. 1987). The geographical distribution of ZCL foci and the propagation of the epidemic in Tunisia were closely conditioned by environmental and ecological factors, including the presence of certain biotopes favourable for rodent hosts, and the development of water resources (dam, wells) and agricultural projects which contributed to an increase in vector population densities (Ben-Ammar et al. 1984; Ben-Ismail and Ben Rachid 1989).

Environmental epidemiology, defined as the study of the spatial or spatio-temporal distribution of disease in relation to possible environmental factors, constitutes an important tool for better understanding the dynamics of parasitic infections and the development of suitable control and prevention strategies (Diggle 1993). GIS is a valuable tool of environmental epidemiology, but has not been extensively used globally in the study of disease distribution and dynamics, although epidemiological data clearly have spatial components. Perhaps the earliest and most celebrated example is John Snow's detection of the Broad Street water-pump as the source of the 1840s cholera outbreak in London. He demonstrated that the association between cholera deaths and contaminated water supplies showed a striking geographical distribution (Diggle 1993; Henk et al. 1991).

[1a]Laboratoire d'Epidemiologie et d'Ecologie Parasitaire, Institut Pasteur, Tunis Belvedere, Tunisia.

[b]Department GIS, Institut Regional des Sciences de l'Information et de Telecommunications (IRSIT), Tunis, Tunisia.

[c]Direction Regionale de La Santé Publique, Sidi-Bouzid, Tunisie.

[d]Direction des Soins de Santé de Base, Ministère de la Santé Publique, Tunis, Tunisie.

In Tunisia, GIS have never been applied to health problems, although there has been geographic recognition of households and manual cartography for improving insecticide coverage within the context of malaria control programs. This approach, although very simple, constituted a valuable tool for the success of malaria control. Unfortunately, this technique was neither improved upon nor replicated for other parasitic diseases such as leishmaniasis and hydatid cystic, or other health hazards heavily influenced by the environment. In fact, this technology could not be sustained manually for two reasons:

- The difficulty of linking multiple and heterogeneous types of data, and
- The overwhelming task of updating dynamic information in the study area.

The advent of computer hardware and software with power and storage capabilities has driven the development of computerized GIS which simultaneously utilize numerous strata of data from a variety of sources. These new technological developments increase the ability of the technology to continuously monitor the pattern of parasitic disease dynamics in Tunisia in both time and space. They also stimulated the introduction of an automated GIS for collecting, storing, retrieving, analyzing, and displaying spatial and nonspatial data regarding parasitic diseases which are highly conditioned by environmental risk factors.

This report presents the preliminary application of an automated GIS for the investigation of the spatio-temporal dynamics of ZCL in relation to potential environmental risk factors, in a pilot focus in Central Tunisia. More precisely, this work aims to:

- Apply the GIS approach in the context of ZCL, to automatically integrate spatial data and attributes related to the disease;
- Display the spatial distribution of cases through time and correlate the occurrence of the disease with environmental risk factors; and
- Refine the calculation of classic epidemetric indices according to the risk level of exposure.

Materials and Methods

The Study Area and Description of Data Collection

The study area covers the "Imadat" (sector) of Felta located in Sidi Bouzid Governorate in Central Tunisia. It is a rural area of 109,850km^2. The primary health care nurse and the sector supervisor carried out a census of the area's total population in 1991, and updated it in 1992 (Ben-Salah et al. unpublished data). Each family in the area was interviewed and their household numbered during door-to-door visits. A pretested questionnaire, used for data collection, contained the following sections:

- Identification of the household:
  - Code of Governorate;
  - Code of Delegation;
  - Code of the "Imadat";
  - Number of the house; and
  - Number of the family
- Description of the dwelling:
  - Type of roof (cement, metal, other);
  - Description of the walls (covered with cement, uncovered);
  - Presence of shelter for animals; and
  - Number of dogs
- Socioeconomic parameters:
  - Profession;
  - Level of parental education;
  - Mode of water supply;
  - Mode of excrement disposal;
  - Availability of electricity and presence of television; and
  - Mode of access to primary health care services
- Family members and other inhabitants:
  - Status in the family: mother, father, son, etc.;
  - Sex;
  - Birth date;
  - School degree; and
  - Occupation
- Information relevant to ZCL:
  - Presence of active lesions or scars;
  - Number of active lesions or scars;
  - sSte of the lesions; and
  - Year and month of the occurrence of ZCL

From 1983 to 1990, the information on ZCL was sought by skin examination performed by local nurses involved in the survey of all individuals in the study area. The face was examined first, then the limbs and the trunk. The diagnosis was in fact very easy, as even the mother or elder relatives could show where the scar was located on the patient. When a typical scar was found, the patient or his/her parents or relatives (if a child) indicated the date of lesion occurrence. Since 1991, information regarding active scars has been collected prospectively through a surveillance system.

Collection of Spatially Referenced Data

Aerial photographs at a scale of 1:20,000 obtained from the Tunisian Office de la Topographie et de la Cartographie have been used by the surveyors during the door-to-door visits. These photographs show the precise location of the dwellings, with a number identifying each house fixed in front. A map (1:20,000 scale) of the entire study area was drawn by a specialist map agency using the aerial photographs.

The constitution of a spatially referenced database was performed using a personal computer to introduce two types of information. Non-locational or descriptive data refer to the features or attributes (variables in the survey form describing individuals, animals or habitats).

Locational or spatial data consist of lines (segments), such as the limits of the area, roads, and so on, points (nodes) corresponding to the dwellings, and areas (polygons) colonized by rodents. Spatial data were digitized into independent coverages of the three types using a format A0 digitizer (Summergraphics) in the GIS department of the Institut Regional des Sciences de l'Information et de Télécommunications (IRSIT, Tunis).

The epidemiological geographic information system (EPI–GIS) was realized using the GIS software package ARC/INFO 3.4D Plus (ESRI, Redlands). Thematic maps were printed using a format A0 plotter and an inkjet colour printer.

The absence of an interface for linking the descriptive database management software (ORACLE PC version) to the GIS software (ARC INFO) required the conversion of the ORACLE data tables into a DBF format compatible with INFO, the database manager of ARC/INFO. Thus, it was possible to join the DBF descriptive data tables to the features attribute tables (AAT: arc attribute table; PAT: point/polygon attribute table) using the command JOIN.

The Epi–GIS conceived and developed by the laboratory of Ecology and Epidemiology of Parasitic Diseases (LEEP) team, permitted the mapping of spatial distribution of ZCL over 10 years and to relate this distribution to the potential vegetation cover, particularly to the presence of chenopods. The command INTERSECT was used to overlay biotope coverages with the dwellings.

Buffers around a focal point believed to be the source of the epidemic were generated with the command BUFFER. This focal point corresponds to a cemetery where *P. obesus* colonies were confirmed in 1983 by field observations (Ben-Ismail et al. unpublished data). Distances between this focal point and dwellings with cases were calculated using the command NEAR.

Data Analysis

The mean force of infection (FOI) for the entire period studied, defined as the number of acts of transmission for one individual occurring within a finite period of time (Lysenko and Beljaev 1987), was calculated for the whole foci using the following formula: $Q(x)= e^{-x}$, where: *Q(x)* is the age-specific proportion of susceptibles (free of scars in the context of ZCL), *x* being the group of age. The analysis involved only the under 10 year-old children in order to control possible age-dependence of the FOI.

The assessment of the FOI for each epidemiologic year was performed from the incidence among susceptibles L, using the following formula: $L(x) = 1-e^{-x}$ (Lysenko and Beljaev 1987), where *x* is the epidemiologic year.

Overlay and intersection operations between different kinds of biotopes (I: presence of chenopods, II: presence of jujubier, III: presence of other plants) and clusters of dwellings located inside these biotopes were performed for the estimation of biotope-wise.

Results

A total of 4,269 individuals belonging to 740 families were interviewed through door-to-door visits. The families lived in 562 dwellings which were numbered during the visits. Since the beginning of the epidemic, 1,081 individuals (22.44%) showed evidence of infection based on the presence of typical scars.

The distribution of the incidence of the infection shows two epidemic peaks in 1985 and 1987. The resulting graph shows neither the pattern of the spatial distribution of ZCL, nor its relation to possible environmental factors.

The distribution of the proportion of susceptibles *Q(x)* over the age groups (0–10 years) revealed a trend for decay. Assuming a constant exponential decay over the age groups (Henk et al. 1991), and using the log likelihood method, *Q(x)* was used to estimate the mean FOI for the whole period of time.

It was found to be 0.017 per year, showing that the focus is hypoendemic for ZCL according to the Lysenko and Beljaev classification (1987).

Using the annual information, the number of immunes (new cases and previous cases), and the number of susceptibles, *L(x)* was estimated and the FOI was directly derived for the corresponding epidemiological year.

Table 1 shows the evolution of the force of infection through time, and the different parameters derived from the database for its estimation. These results show that the FOI, far from being stable over time, follow the same trend as the incidence of infection with two peaks in 1985 and 1987.

Table 1. Evolution of the Force of Infection Through Time and the Parameters Used for Its Estimation.

Year	Susceptibles	ZCL Cases	Incidence L	Force of infection λ
1983	3334	15	0.004	0.005
1984	3444	64	0.018	0.019
1985	3489	392	1.112	0.119*
1986	3258	53	0.099	0.016
1987	3351	332	0.099	0.104*
1988	3142	69	0.021	0.022
1989	3183	73	0.022	0.023
1990	3215	21	0.006	0.007
1991	3315	37	0.011	0.011
1992	3486	24	0.006	0.007

*Values of λ corresponding to epidemic peaks in 1985 and 1987.

To visualize the spatial dynamics of the disease, an overlay between the base map of the area and the coverage of the dwellings without cases was achieved (Fig. 1a). The different biotopes encountered in the study area are displayed in Fig. 1b.

Figure 2a,b,c,d, presents maps displaying the spatial distribution of dwellings with ZCL cases, for the years 1983, 1985, 1987, and 1992, respectively. The location of the first cases which occurred in 1983 (Fig. 2a), allowed the identification of the initial starting point of the epidemic in a cemetery that had been highly colonized by the *Psammomys obesus*, the main reservoir of the parasite.

The capabilities of the EPI–GIS were applied to select clusters of dwellings according to the kind of biotope where they were located. This procedure permitted the comparison of the FOI in different biotopes during the epidemic years of 1985 and 1987 (Table 2). The results showed a large heterogeneity of the force of infection in the different selected biotopes, ranging from 0.23 in the biotope of chenopods to 0.01 in other biotopes in 1985, and confirmed that the distribution of ZCL is very sensitive to environmental factors.

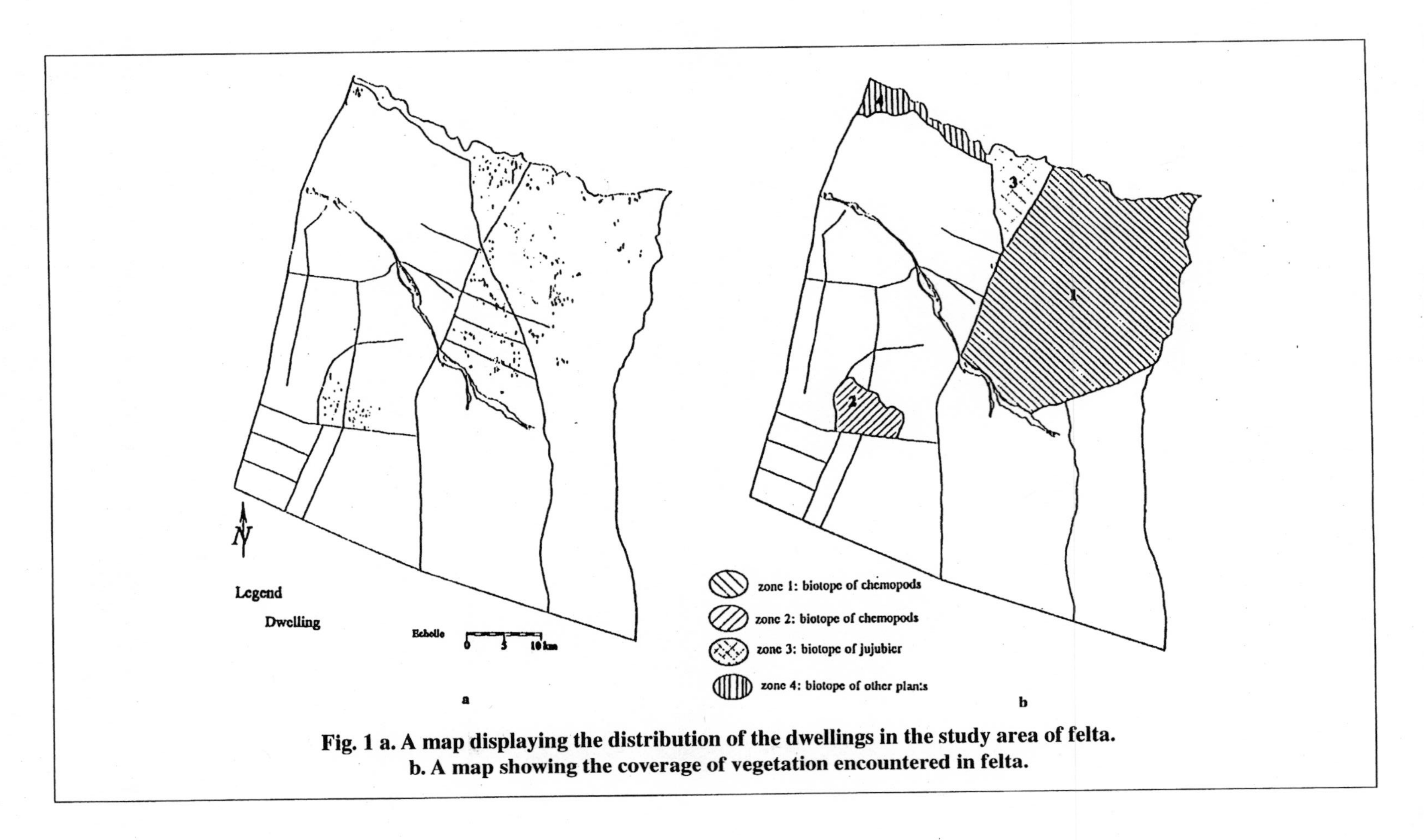

Fig. 1 a. A map displaying the distribution of the dwellings in the study area of felta.
b. A map showing the coverage of vegetation encountered in felta.

Fig. 2a,b,c,d. Distribution of maximum, minimum, and mean distances between the cemetery constituting the starting point of the epidemic and the dwellings with ZCL cases through time.

a. **Location of the dwellings with the first cases of ZCL epidemic in 1983 near an area highly colonized by *Psammomys obesus* the main reser voir of the infection.**
b. **Spread of ZCL to the surrounding dwellings during the epidemic year of 1985.**
c. **Extension of the epidemic to the dwellings located in the north-east of the area during the epidemic year of 1987.**
d. **Distribution of dwellings with sporadic cases in 1992.**

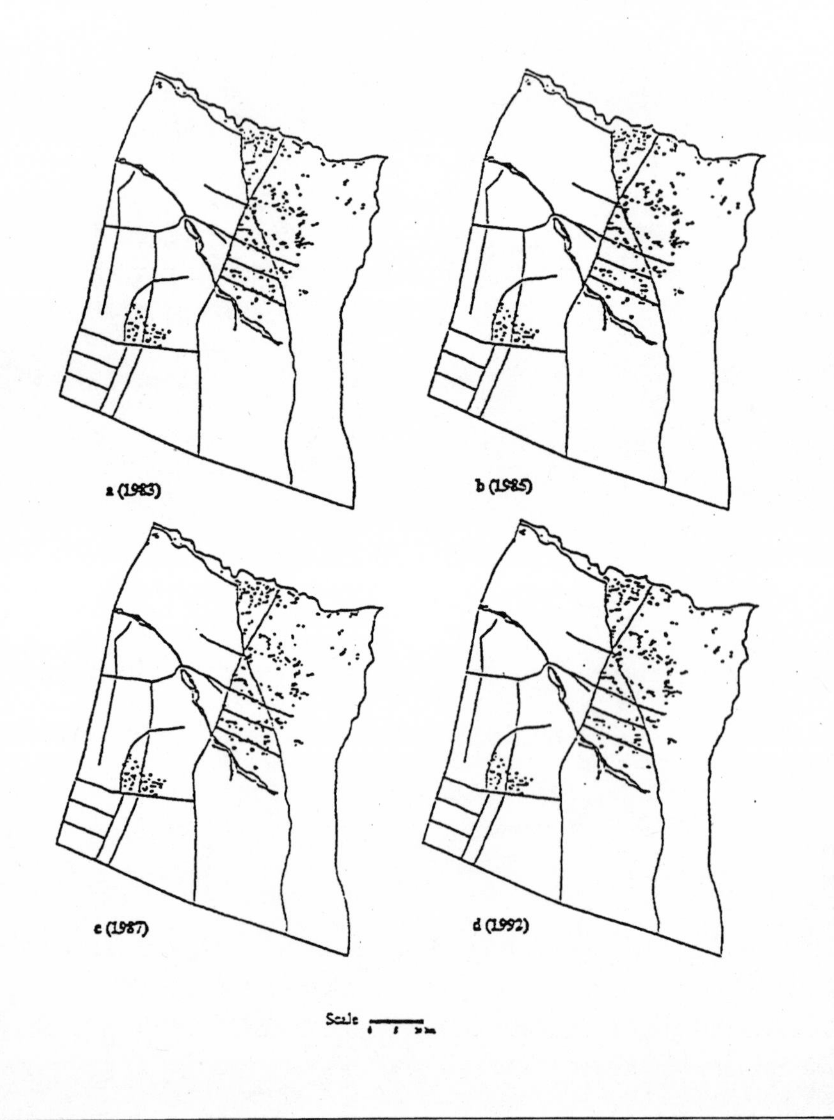

Table 2. Estimation of the Force of Infection λ in Different Biotopes during the Epidemic Years (1985, 1987).

Type of Biotope	Population Size	Susceptibles	ZCL cases	Incidence	Force of infection λ	Level of endemicity *
1985						
1	2578	2126	239	0.112	0.119	méso
2	373	340	71	0.209	0.234	méso
3	656	635	23	0.036	0.037	hypo
4	201	189	3	0.016	0.016	hypo
1987						
1	2373	2309	354	0.153	0.166	méso
2	346	346	28	0.081	0.084	méso
3	606	597	1	0.002	0.002	hypo
4	185	185	9	0.049	0.05	hypo

1 and 2: biotope of chenopods with borrows of the rodent *Psammomys Obesus* the main reservoir of ZCL.
3: biotope of jujubier with borrows of the rodent Meriones shawi secondary reservoir of ZCL.
4: other kinds of plantation having no relation with the reservoir of ZCL.
*: the level of endemicity is rated according to Lysenko and Beljaev classification.

Indeed, in the biotopes 1 and 2 of chenopods the population of rodents is very dense, and individuals are exposed to a high risk of transmission. In these biotopes, the infection is mesoendemic, while it is hypoendemic in biotope 3, where the transmission is expected to be lower due to the absence of the infection reservoir. This analysis confirmed the heterogeneity of the transmission within the same small focus.

Buffers were generated every 200m around the cemetery at the beginning of the epidemic in 1983, to estimate the evolution of the FOI around this starting point.

The distances between the starting point of the epidemic and the dwellings with cases ranged between 251m and 1,102m in 1983. The maximum distance increased very quickly in 1984 to reach 2951m, and remained almost unchanged over the following years.

Discussion

A retrospective survey undertaken in the endemic sector of felta permitted the estimation of the crude prevalence of ZCL and the distribution of incidence through time. It clearly demonstrated the importance of ZCL as a public health problem in the study area and the need for control measures.

The calculation of such classical epidemiological indicators usually includes the whole population surrounded by administrative boundaries of the district for deriving the indicator denominator (incidence, prevalence), assuming that all individuals are under an equal risk of infection. This assumption, although plausible for many health problems, is clearly misleading for vector-born zoonosis where the abundance of both the vector and the reservoir is highly dependant upon the type of biotope where people are living. Within the context of ZCL, the EPI-GIS combined demographic, medical, and geographic data, and allowed the estimation of FOI through time and space. It also facilitated the study of relationships between the distribution and diffusion of the disease, and environmental risk factors. It offered the opportunity to test the plausibility of the basic assumptions of Lysenko and Beljaev's model (the homogeneity of the transmission through time and space) (1987). It was clearly demonstrated that these assumptions are unrealistic in the Tunisian context, raising the need for considering spatial heterogeneities and instability through time in the estimation of the force of infection.

GIS clearly confirmed the role of chenopods as determinants of transmission, and permitted the display of precise spatial distribution of infection. This result corroborates previous findings based on different epidemiological approaches (Ben-Salah et al. 1994). GIS allowed the refinement of epidemiological parameters and the stratification of the studied area to different levels of risks for the transmission, which has very important implications in terms of understanding the determinants of the infection, its dynamics, and the allocation of resources for control.

By evaluating the distances between the source of the epidemic and the dwellings with cases, the reach of an epidemic of ZCL is estimated as approximately one kilometre. Such information is extremely important for designing control measures based on environmental changes, such as plowing the zones of chenopods. It informs precisely on the minimum area necessary to be ploughed to interrupt the transmission of the infection to people.

This approach appears to be more realistic than one based on the measurement of sandfly vector dispersal (maximum flight distance) (Killick–Kendrix et al. 1984). This research stimulated the collection of other environmental factors such as soil type, climatological parameters, waterways, and proximity to parasite sources to establish the relationships between these parameters and the spatial patterns of the disease.

In the context of ZCL, EPI–GIS provides a excellent framework within which the overlaying of a large number of geographical factors (coverages) and the association of them with the disease, the vector, and the reservoir can be undertaken. The GIS approach will become more important if the present research is extended to a larger geographical region. As more data on the trophic preferences and dynamic of the reservoir population become available, the EPI–GIS can assist epidemiologists to predict the spread and establishment of the disease in new locations which constitute a valuable tool for resource allocation and the implementation of preventive measures.

Acknowledgments

This investigation received financial support from the UNDP/World Bank/WHO Special Programme for Research and Training in Tropical Diseases (TDR/RSG, Grant: ID: 890266), from the "Secrétariat d'Etat à la Recherche Scientifique et à la Technologie de la Tunisie" (Projet DRDT RF 2170), and from IDRC, Canada (Project Ref. 92–0233).

References

Ben-Ammar, R.; Ben-Ismail, R.; Helal, H.; Bach-Hamba, D..; Chaouch, A.; Bouden, L.; Hanachi, A.; Zemzari, A.; Ben-Rachid, M.S. 1984. Un nouveau foyer de leishmaniose cutanee de type rural dans la region de Sidi-Saad, Tunisie. *Bull. Soc. Fr. Parasitol.*, 2, 9–12.

Ben-Ismail, R.; Ben-Rachid, M.S. 1989. Epidemiologie des leishmanioses en Tunisie. In *Maladies Tropicales Transmissibles*. AUPELF-UREF John Libbey, ed., Eurotext, Paris, France. pp. 73–80.

Ben-Ismail, R.; Gramiccia, M.; Gradoni, L.; Ben-Said, M.; Ben-Rachid, M.S. 1987. Identificazione biochemica di isolati leishmania della Tunisia. *Parassitologia, 28,* 186–187.

Ben-Ismail, R.; Gramiccia, M.; Gradoni, L.; Helal, H,; Ben-Rachid, M.S. 1987. Isolation of Leishmania major from *Phlebotomus papatasi* in Tunisia. *Trans Roy. Soc. Trop. Med. Hyg.*, 81, 749.

Ben-Ismail, R.; Ben-Rachid, M.S.; Gradoni, L.; Gramiccia, M.; Helal, H.; Bach Hamba, D. 1987. La leishmaniose cutanée zoonotique en Tunisie. Etude du réservoir dans le foyer de Douara. *Ann. Soc. Belge Med. Trop.* , 67, 335–343.

Ben-Ismail, R.; Ben-Salah, A.; Hidoussi, A.; Iammel, J.; Hellal, H.; Ben-Ammar, R.; Sidhom, M.; Dellagi, K. 1991. Apport de la cartographie aérienne dans le recuil des données épidémiologiques de base au cours des zoonoses: un exemple dans un foyer pilote de Kala azar. *7ème congrés Médical Arabe/ 1er congrés Médical Tunisien*. Tunis 18, 19, 20 Octobre 1991.

Ben-Salah, A.; Mbarki, L.; Ftaiti, A.; Balma, A.; Chahed, M.K.; Garraoui, A.; Sidhom M.; Dellagi, K.; Ben-Ismail, R. 1994. The prevalence and risk factors associated with zoonotic cutaneous leishmaniasis in central Tunisia. Abstracts of the 8th International Congress of Parasitology. 10-14 October, Izmir, Turkey, 189.

Diggle, P.D. 1993. Point process modelling in environmental epidemiology. In Statistics for the environment. Vic B.; ed., Turkman K. F.; Wiley, London, UK.

Henk, J.; Scholten, M.; de Lepper, J.C. 1991. The benefits of the application of geographical information systems in public and environment health. *World health statist. quart.*, 44, 160–170.

Killick–Kendrick, R.; Rioux, J-A.; Bailly, M.; Guy, M.W.; Wilkes, T.J.; Guy, F.M.; Davidson, I.; Knechtli, R.; Ward, R.D.; Guilvard, E.; Perieres, J.; Dubois, H. 1984. Ecology of leishmaniasis in the south of France. 20. Dispersal of Phlebotomus ariasi Tonnoir, 1921 as a factor in the spread of visceral leishmaniasis in the Cévennes. *Ann. Parasit. Hum. Comp.*, 59, 123–156.

Lysenko, A.J.; Beljaev, A.E. 1987. Quantitative approaches to epidemiology. In The leishmaniases in biology and medicine. Peters, W. and Killik–Kendrick B., ed. Academic Press, London, UK. pp. 263–290

Use of RAISON for Rural Drinking Water Sources Management

C.W. Wang[1]

Introduction

The safety of rural water supplies is a critical issue in developing countries. Heterogenous rural water sources can pose serious health risks for users unless timely testing, reporting, and record-keeping of water-borne diseases and likelihood of contamination are controlled.

The RAISON GIS (regional analysis by intelligent systems on microcomputers geographical information system) was designed to manage the microbiological quality of rural drinking water sources, by providing an efficient means for the storage, analysis, and presentation of a large volume of water quality monitoring data.

The International Development Research Centre (IDRC) sponsored the development of this information system at the National Water Research Institute in Burlington, Ontario, Canada (Tong et al. 1989), for the purpose of data collection and processing in and by both developed and emerging nations. First used to assess the risks posed by acid rain to water sources in Eastern Canada (Lam 1986), the RAISON system is an integrated package incorporating a database management system, a mapping package, and a spreadsheet. It utilizes a set of GIS programs under the RAISON Programming Language environment for a multilevel, menu-driven presentation of water quality data and related information in map forms.

Based on the RAISON system, a prototype system called µRAISON was developed at the University of Malaya as part of a rural water supply, sanitation, and drinking water surveillance program. µRAISON uses additional common software packages for certain parts of its implementation, such as dBASE III for complex database management, Autocad or Crosstalk for map digitization, and SPSS PC+ for statistical analysis. A separate databases management package was also employed in view of the versatility of dBASE III.

[1]Department of Biochemistry, Faculty of Medicine, University of Malaya, Kuala Lumpur, Malaysia.

μRAISON System Design and Configuration

Principal Tasks

To set up the μRAISON system, digitized map files must be obtained, showing maps at the national, state, district, and village levels. Map files can be digitized using a table-top digitizer with the aid of other software packages such as Cross-talk or Autocad. After the necessary conversion and edition, these map files are incorporated into μRAISON. Water sources or sampling stations are represented at the village level, and are identified by station names and specific sets of longitude-latitude readings. These readings are processed in a format specific to μRAISON. For example, a reading of 23°59, which is equivalent to 23.999°, is entered as 23999. The longitude values are measured from west to east.

The second task involves the creation of various database files, in which geographical locations are identified by state, district, village, or station names. Statistical information on water supply, sanitation facilities, and other relevant socioeconomic information for the respective geographical levels is entered directly into the RAISON database. However, to facilitate a more direct access for statistical analysis, water quality monitoring data and other complex sanitary survey features have been set up in dBASE III.

The third task is to develop the programs in the RAISON Programming Language environment so as to allow interactive menu-driven data analysis and presentation. Complex steps involving database and spreadsheet operations linking with GIS presentations are transparent to the users.

μRAISON System Configuration

The setting up tasks of the μRAISON Water Quality Data Management System involve map digitization and the creation of database files. Updating could be input externally. However, statistical analysis is carried out separately from the μRAISON package.

Map digitization uses packages such as Cross-Talk or Autocad for the generation of vector graphic files. The files are then converted into map files for incorporation into RAISON using the "calibration/conversion programs" developed for μRAISON.

Water quality and sanitary survey databases are created in the dBASE III+ environment and then exported to the main RAISON database for source classification and map presentation. Within the μRAISON system, procedures have also been developed for exporting these database files to SPSS PC+ for statistical analysis.

Two additional options are also available for μRAISON after system startup, namely :

- Accessing the general operational commands of RAISON through the main menu, and
- Performing specific water quality data management tasks using the geographical information system.

By way of the main menu, users can invoke the various map icon files, map snapshots and geographical information displays in an unguided manner. The GIS option is a multilevel, menu-driven system for performing specific tasks in water quality data management for rural water supply. The basis of the multilevel structure is the local village level (level 2), at which individual water sources for each house-hold are monitored. This information is then compiled and analyzed at the levels of district (level 1B), state (level 1A) and national (level 0).

μRAISON Users

The μRAISON system was targeted for two levels of users. Using the menu-driven GIS options, an operator with minimum computer knowledge can update data inputs and execute specific data analysis and GIS presentations. Advanced users can follow the user manual to add or change functions in the data analysis and map presentation to meet their specific needs for management and planning.

Sanitary Survey and Water Quality Data

More than 1,500 pieces of field testing data on the microbial water quality of some 800 water sources have been compiled. These have been stored in dBASE III+ files and imported into the μRAISON database to be utilized for testing, programing for analysis, and display within the μRAISON system environment. A total of 15 village/region map files have been incorporated. These include four villages each in the States of Selangor and Negeri Sembilan, one village in the State of Pahang, three regions in the State of Kedah, and three villages in the State of Kelantan.

Sanitary survey records for the sampling stations were also obtained at the time of sampling, and a database file designed for their storage. The database file consists of four sections: type of water supply, sanitary protection of the water source, sources of pollution, and land usage.

The data gathered on water supplies and sanitary surveys provided such information as: type of well (e.g., dug, tube, pipe, public water supply, open watercourse), state of repair (as evidenced by cracking or crumbling casing or presence of rubbish), surrounding ground type (clay, sand, etc.), population density, presence of small children, and agricultural activity (particularly the presence of animals). In addition, data on type, availability, and status of adjacent latrines were included. These data are relevant as many examples of well contamination can be related to the presence of latrines which are inefficient due to their flushing system, the large number of individuals using them, their proximity to wells or gradient from wells (upslope or downslope), and their general state of repair/cleanliness.

Water Quality Classification System

The analysis of water quality data within the GIS will depend on the classification system designed by the rural water sources. At present, there are no established classification models for rural drinking water sources. The basic approaches of various classification systems which have been adopted or developed are summarised as follows:

Subjective Classification System Based on Commonly Adopted Standards

Faecal coliforms have been widely accepted as the indicator organisms of choice for the detection of human faecal contamination of water sources. Wang et al. (1989) have also demonstrated the usefulness of coliphage enumeration for this purpose.

Data obtained on coliphage and m-FC (membrane faecal coliforms) counts have been used directly for the classification of the rural water sources. Ranges of coliphage or m-FC counts of <5, 5–50, 50–250, 250–1,000, and >1,000 per 100 ml sample have been classified as Class I, II, III, IV, and V, respectively. A systematic colour code has been adopted in the μRAISON GIS for the display of water sources quality in map format. Class I quality is consistent with the WHO guidelines for drinking water (1984). Classes II and III represent moderate quality in which water requires pretreatment such as boiling or chlorination before drinking. Classes IV and V are poor quality water of high risks which called for remedial actions on the water sources.

The usefulness of this classification scheme can be judged on the basis of the data sets tested. The classification results obtained for both m-FC and coliphage data for the 1,500 water sources in 15 villages/regions show excellent consistency.

Objective Classification Using a Single Parameter

An objective classification system for demonstrating the quality of rural water sources can be obtained using a single parameter which follows a negative binomial distribution based on the estimation of the index of dispersion for the observed data (Shaarawi et al. 1981). The coliphage or m-FC has been chosen as the single parameter to be classified separately .

The analytical results showed that if data on very high coliphage counts (>240 counts/20 ml or 1,200 counts/100 ml) were not included in the fitting, the coliphage data follows a negative binomial distribution.

The group classification obtained based on coliphage counts was as follows:

	Counts/20 ml	Counts/100 ml
Class I	<1	<5
Class II	>1–10	>5–50
Class III	>10–31	>50–155
Class IV	>31–62	>151–310
Class V	>62	>310

More classes were derived if data of higher coliphage counts were included, although the ranges of lower groups were not affected. It is reasonable to group all data values >62 counts/20 ml (>310 counts/100 ml) as the highest group, due to the small number of samples observed over the range. It is noted that the classification results are comparable to those obtained using the subjective method.

Ranking Method Using Several Parameters

A ranking procedure taking microbial counts and sanitary protection conditions of water sources into account has been incorporated into the GIS options. The sanitary ranking for each water source has taken into account the source protection, proximity of polluting sources, and soil type.

Preliminary statistical correlational tests were performed between the microbial counts and the sanitary protection conditions, using SPSS PC+ to define the ranking for water sources of high quality (class I) to high risk (class V). Statistical analysis was carried out for (a) type of well, (b) depth of well, and (c) well protection, versus the different ranges of m-FC or coliphage counts. All six cases showed significant correlation at the 0.1% level between the microbial counts and the sanitary conditions.

Multivariate Classification Based on Discriminant Analysis

Tong et al. (1990) have presented a multivariate classification scheme based on the discriminant analysis for rural potable water sources. The multivariate classification approach is based on discriminant analysis which correlates sanitary protection, polluting sources, land use information, as well as field data on microbiological quality. Tong et al. have shown that a linear combination model using the sanitary condition data correctly classified 78% and 73% respectively of the m-FC and coliphage data into two distinct groups each (<5 and >5 counts/100 ml). The corresponding percentage of correct classification was 70% when the m-FC counts were divided into three groups (<15, 15–>500, >500 counts/100 ml). Similarly, the corresponding percentage of correct classification was 66% when the coliphage counts were considered in three groups (<5, 5–<250 and >250 counts/100 ml).

References

El-Shaarawi, A.H.; Esterby, S.R.; Dutka, B.J. 1981. Bacterial density in water determined by Poisson or negative binomial distributions. *Appl. Environ. Microbiol.*, 41, 107–116.

Lam, D.C.L. 1986. Computer simulation of watershed acidification. Proceedings, Summer Computer Simulation Conference. Reno, USA. Soc. Computer Simulation, San Diego, CA, USA. pp. 456–460.

Tong, S.L.; Lim, C.K.; Wang, C.W.; Chin, S.T.; Loh, C.L.; Lim, K.C.; Ho, P.Y.C.; Lam, D.; El-Shaarawi, A.; Swayne, D.; Storey, J. 1989. µRAISON user manual. University of Malaya, Kuala Lumpur, MY.

Tong, S.L.; Wang, C.W.; Ho, Y.C. 1990. Classification systems for rural potable water sources. Proceedings Second International Biennial Water Quality Symposium: Microbiological aspects. Castillo et al., ed. pp. 312–323.

Wang, C.W.; Ngeow, Y.F.; Loh, C.L.; HO, Y.C. 1989. Water quality control. Final report, University of Malaya, MY.

WHO. 1984. Guidelines for drinking water quality. World Health Organization, Geneva, Switzerland.

Appendicies

Interests, Problems, and Needs of GIS Users in Health: Results of a Small Survey

Luc Loslier[1]

A survey of participants was taken during the workshop to better determine their interests and needs regarding GIS applications and functions. This paper serves to bring forward the salient points from that survey. It is hoped that the results will facilitate the development of better solutions to the problems encountered by the growing community of GIS users in the health field.

Survey responses are presented in Table 1. Responses range from I (very useful) to V (not useful) for questions 1, 2, and 3; and Yes or No for question 4. For each variable, the highest frequency has been printed in bold. Question 5 was open-ended, and will be commented on in the conclusion.

Questions 1 Through 4

Question 1 dealt with the usefulness of GIS in different health-related fields. GIS are seen as most useful for monitoring and control, which was given a score of I by 12 respondents, and a score of II by 7 respondents. GIS were also believed to be very useful for research, with 85% of the respondents rating their usefulness as I or II. Responses varied more across health policy: 12 respondents gave it a score of I or II, while 7 gave it a score of III. Finally, the usefulness of GIS for health education was not seen to be as great by many of the participants: only eight persons gave them a score of I or II.

Question 2 looked at the geographic scale of GIS applications, ranging from the local to the national level. The level at which GIS were seen to be the most useful was the national. However, if scores of I and II are considered together, all levels are considered relatively equally: 14 respondants chose the national level as that at which GIS would be most useful, 14 also chose the district / large urban settings level, and 15 chose the regional/provincial level.

Question 3 was more technical and dealt with GIS operations. This question was characterised by a great homogeneity of responses. Not a single GIS operation, among the 28 listed in the question, was rated poorly. At the same time,

[1]Université du Québec à Montréal, Montréal, Québec, Canada.

Table 1. Results of GIS Workshop Survey.

Variable	*Variable Label*	*Category*				
		Very Important				*Not Important*
		I	II	III	IV	V
	Usefulness of GIS in field of:					
Q1EDUCAT	Health Education	2[2]	**6**	5	4	3
Q1MONITR	Health Monitoring and Control	**12**	7	0	0	0
Q1OTHER	Usefulness - other	**5**	1	1	0	3
Q1POLICY	Health Policy	**8**	4	7	1	0
Q1RESRCH	Health Research	8	**9**	2	1	0
	Usefulness of GIS at level of:					
Q2RURAL	Small community / Rural setting	4	**6**	3	4	1
Q2DISTRC	District / Large urban setting	**7**	**7**	3	1	0
Q2REGION	Region / Province	7	**8**	2	0	1
Q2NATION	Nation	**11**	3	1	3	1
	Importance of GIS operations:					
Q31COLLC	Spatial data collection	0	4	**15**	0	0
Q31SCAN	Spatial data scanning/editing	2	8	**10**	0	0
Q31DIGI	Digitizing	0	3	**17**	0	0
Q31SATEL	Satellite images input	6	3	**9**	0	0
Q31RASTR	Raster - Vector conversion	2	5	**10**	0	0
Q31CARTP	Cartographic projection transform.	2	4	**11**	0	0
Q31DOCUM	Spatial database documentation	2	2	**14**	0	0
Q32ATCOL	Attribute data collection	1	3	**14**	0	0

[2]This represents the number of people answering with this response.

Q32DESGN	Attrib. database design,docum.	2	5	**13**	0	0
Q32ENTRY	Attrib. data entry, transform.	2	5	**13**	0	0
Q32QUERY	Attrib. data queries	3	3	**12**	0	0
Q32STAT	Attrib. data statistical analysis	1	4	**13**	0	0
Q32REPRT	Producing reports	1	6	**12**	0	0
Q33OVERL	Map overlay	0	3	**17**	0	0
Q33ALGEB	Map algebra	3	**6**	**6**	0	0
Q33CONTR	Interpolation and contouring	2	4	**10**	0	0
Q33ELEVA	Digital elevation modeling	4	**8**	4	0	0
Q33BUFFR	Buffer zones construction	2	2	**15**	0	0
Q33COST	Cost surface and Path analysis	2	4	**9**	0	0
Q33LOCAL	Location allocation modeling	1	3	**12**	0	0
Q34MTYPE	Types of maps selection	1	6	**13**	0	0
Q34MBASE	Selecting base maps	2	4	**14**	0	0
Q34MDESI	Map design	2	**9**	8	0	0
Q34MTHEM	Thematic mapping	4	3	**10**	0	0
Q34MOUTP	Mapping output	0	8	**11**	0	0
Q35PROJE	Project description	1	2	**12**	0	0
Q35CONTA	Contact supply target groups	4	5	**10**	0	0
Q35RESUL	Results diffusion	2	7	**9**	0	0
	Avail.of skilled personnel:	*No*	*Yes*		*Ctry*	*Inst*
Q41SPMAN	Spatial data magmt pers.	7	**13**		**9**	3
Q41ATMAN	Attrib.data magmt pers.	6	**14**		**7**	5
Q41MODEL	Spatial modeling skilled pers.	**16**	4		**12**	1
Q41MAPPG	Mapping skilled pers.	9	**11**		**9**	3
Q41PROJC	Project planning Managmt skilled pers.	4	**16**		4	**7**

INTERNATIONAL WORKSHOP FOR HEALTH AND THE ENVIRONMENT

5 - 10 September, Colombo, Sri Lanka

GIS for Health - Quick Workshop Survey

To assist identification of directions for follow-up of this Workshop, would you kindly answer the following questions, based on your former GIS working experience and what has emerged from the Workshop.

(Please answer questions 1 and 2 using a number between 1 (very useful) and 5 (not useful).

1. What is your opinion about the usefulness (1-5) of applying GIS

   a) in Health Research ☐

   b) in Health Monitoring and control ☐

   c) in Health Policy ☐

   d) in Health Education ☐

   e) Other (specify) ☐

2. What degree of usefulness (1-5) would a Health GIS have when applied at the following levels

   a) Small community I Rural setting ☐

   b) District / Large Urban setting ☐

   c) Region / Province ☐

   d) Nation ☐

3. Taking into account your work setting and conditions, and your specific needs and problems, please indicate on page 2 the relative importance of the listed GIS operations. If you think that some operations should be added, please do so on a blank line.

4. The main tasks for operating a GIS are listed below (for details of these tasks please refer to table on page 2). Adequately trained personnel are necessary to realize these tasks.

4.1 Do you have personnel with appropriate skills for : (Y or N)

   a) Spatial Database Management ☐

   b) Attribute Database Management ☐

   c) Spatial Modeling ☐

   d) Mapping ☐

   e) Project Planning and Management ☐

4.2 If no, do you have access to people with these skills elsewhere in your institution or in your country?

		COUNTRY	INSTITUTION
a)	Spatial Database Management	☐	☐
b)	Attribute Database Management	☐	☐
c)	Spatial Modeling	☐	☐
d)	Mapping	☐	☐
e)	Project Planning and Management	☐	☐

5. It has been suggested that a network of people interested in the Health field be created following this Workshop.

Could you express your idea about how this network could be of mutual benefit? Do you have suggestions about how it should function, what it should offer?

Will you please indicate your name :______________________________

Organization:____

Address:____

Telephone:____________ Fax :____________ Email address: ____________

Thank you for answering these questions. Could you please le ave our filled questionnaire at The Workshop Secretariat (Sapphire Room) at the latest Friday 9, 2 o'clock.

GIS in Health — Importance of GIS operations in a health perspective

GIS Operation	Very important	Important	Medium importance	Not important
1. Spatial database management				
Data collection I type and sources — raster, vector; ma s, satellites				
Scanning , editing				
Digitizing				
Satellite images input				
Raster - vector conversion				
Cartogra hic projection transformation				
Database documentation				
2. Attribute database management				
Data collection I type and sources —Census, surve ; rimar , secondar				
Database desi n and documentation				
Data entr , editin , transformation				
Data queries				
Statistical anlysis				
Producin reports				
3. Spatial modeling				
Map overlay				
Map algebra				
Intepolation and contouring				
Digital elevation modeling				
Buffer zones construction				
Cost surface and path analysis				
Location allocation modeling				
4. Mapping				
Types of maps				
Selectin base maps				
Map design				
Thematic mapping				
Mapping output				
5. Project planning and management				
Project description — Objectives, methods, data, schedulin , budget				
Contacts - suppliers, managers, target groups				
Results diffusion				

however, few GIS operations were rated as "very important". Rating III was themost common answer, a result which is examined below. It is interesting to note that satellite images input received the highest number of 'very important' scores (6).

Question 4 considered the human resources required to operate a GIS, a complex technology which requires skilled personnel to realize its full potential. Most participants indicated that they have skilled personnel for all the tasks listed, except spatial modelling, but there were also a significant number of persons who do not have the required personnel.

Project planning and management is the least problematic, with only four persons not having skilled personnel. Database management is more problematic: six respondents face a lack of personnel in attribute database management, while seven respondents have personnel problems in spatial database management. Mapping brings more difficulties: almost half of the respondents do not have personnel skilled in this area. The most important problem is with spatial modelling: nearly all respondents (85 %) lack skilled personnel in this area.

Question 4.2 was prepared for the respondents who answered 'No' in the preceding question. However, the data shows that not only these people responded: the number of responses in question 4.2 exceeds the number of persons who answered "No" in question 4.1.

This makes the data rather ambiguous and difficult to interpret, but it nevertheless permits an important observation: many persons who do not have skilled personnel for some GIS tasks in their immediate environment can find them elsewhere within their institution. This is only part of a solution, however, as there remain clear problems at the institutional level with attribute data management (one case), spatial data management (four cases), mapping (six cases), and spatial modelling (15 cases). Most respondants indicated that if they could not find skilled GIS personnel within their own institution, they could them elsewhere in the country, except for persons skilled in spatial modelling, which still causes problems in four cases.

Discussion

There have been attempts to identify a few "key patterns" in the survey data presented in this text, using methods such as nonlinear principal components analysis and cluster analysis. The hypothesis was that different kind of GIS users (for instance users from universities versus users from ministries, or research-oriented users versus application-oriented users) would have different profiles. These attempts to pattern the responses were unsuccessful, however, because the variability of the data is very small. In other words, the population of GIS users investigated seems to be very homogeneous; in general, they share the same vision, interests, and difficulties. A few points, such as the use of GIS for health education, or their application at the community or rural level, show a wider variation of opinions. This general homogeneity, as well as the divergence on these specific points, are worth of consideration.

A second issue which should be noted is that the "application questions" (questions 1 and 2) clearly received better scores than the more technical items of question 3. This probably means that current users of GIS in the health field are more concerned with the application side of the technology and less concerned by its more technical aspects.

These are not seen as not important (columns IV and V have frequencies of 0 for all GIS functions!), but they clearly have fewer high ratings than application questions. The exceptions should be noted: map algebra ("reasoning," i.e., following a logical or mathematical path with maps), digital elevation modelling (the importance of the physical environment could be stressed here), and map design are seen as specially important among the GIS functions.

The responses concerning the availability of skilled personnel to assist in the operation of a health-oriented GIS indicate that many participants have problems at this level: the answers indicate that 42% of personnel needs remain unmet. It should be stressed that the GIS operations for which the unmet needs are the greatest (spatial modelling, mapping, and spatial database management) are precisely at the heart of a GIS, and are probably its most characteristic functions.

The survey data presented indicates that to exploit to the full potential of GIS, health specialists need a system that offers a wide variety of "user friendly" operations. Specifically, a system is needed that offers the required tools but that allows the user to concentrate on application problems. More could probably be done to devise such a system, perhaps by adapting existing packages. The data concerning the availability of skilled personnel leads to the same conclusion: an effort should be made to better adapt the spatial components of the GIS to the health field.

Conclusions

Increasingly more people believe that enormous progress in public health can be made by better understanding the links between health and the human and physical environment. GIS are certainly among the most useful tools that make possible the analysis and representation of these complex relations. Users of GIS in the health community still appear as pioneers; their experience shows that the use of GIS is not without difficulties and that the tool could probably be better adapted to their situation.

The survey presented here indicates that if the tool is to be used to its full potential, efforts should focus on several issues.

- Developing a software system that offers a range of GIS functions in a more user-friendly way to people who are not "hard GIS specialists." The ideal Health GIS program should let the user concentrate on the substance of the problem rather than on the tool.
- Finding a solution to the scarcity of skilled personnel. Development of a system as described in the preceding paragraph should help alleviate this problem. The preparation of a 'health-oriented tutorial package' to accompany such a system, presenting examples of health/environment data integration, analysis, and presentation could also help solve this problem.
- Increasing accessibility of the spatial data. There is a wealth of remotely sensed environmental data (climate, relief, vegetation, land use, transport networks, human habitat, to name a few), but where is that material? How is it obtained? How much does it cost? This seems to be far from clear for health specialists. Some kind of information centre could be imagined to facilitate the access of health specialists to this data.

This last idea brings us back to the survey questionnaire and question 5, which was an open ended question concerning the creation of a network for Health GIS. A strong need was expressed by each respondant for reducing isolation and fostering a constant exchange of information about problems, experiences, methodology, tools, software, and so on. Many suggestions were made to augment the interaction among the health GIS users: newsletters, journals, meetings with presentation of papers, workshops, clearing houses, bulletin boards, and use of the Internet (although not everyone has access). The form and content of this Health/GIS forum remains to be more clearly defined, but it seems to be worth doing and unanimously awaited.

GIS, Health, and Epidemiology: An Annotated Resource Guide

Steven Reader

This guide is divided into four sections: GIS, health, and epidemiology references; GIS books, journals, and magazines; GIS software; and GIS internet resources. This guide is not intended to be comprehensive but may provide some useful starting points for anyone interested in applying GIS to the health and epidemiology fields.

GIS, Health, and Epidemiology: Selected References

(Note: These references are taken from a list compiled by Wayne Hall at St Marys University, Halifax, Canada. Wayne can be reached at b-hall@bass.stmarys.ca).

Beck, L.R. et al. 1994. Remote sensing as a landscape epidemiologic tool to identify villages at high risk for malaria transmission. *American Journal of Tropical Medicine & Hygiene,* 51(3), 271–280.

A landscape approach using remote sensing and geographic information system (GIS) technologies was developed to discriminate between villages at high and low risk for malaria transmission, as defined by adult *Anopheles albimanus* abundance. Satellite data for an area in southern Chiapas, Mexico were digitally processed to generate a map of landscape elements. The GIS processes were used to determine the proportion of mapped landscape elements surrounding 40 villages where *An. albimanus* abundance data had been collected.

The relationships between vector abundance and landscape element proportions were investigated using stepwise discriminant analysis and stepwise linear regression. Both analyses indicated that the most important landscape elements in terms of explaining vector abundance were transitional swamp and unmanaged pasture. (Abstract truncated).

Gesler, W. 1986. The uses of spatial analysis in medical geography: A review. *Social Science & Medicine*, 23(10), 963–973.

This paper is a review of how geographers and others have used spatial analysis to study disease and health care delivery patterns. Point, line, area and surface patterns, as well as map comparisons and relative spaces are discussed. Problems encountered in applying spatial analytic techniques in medical geography are pointed out. The paper is intended to stimulate discussion about where medical geography can and should go in this area of study. Point pattern techniques include standard distance, standard deviational ellipses, gradient analysis and space and space-time clustering. Line methods include random walks, vectors and graph theory or network analysis. Under areas, location quotients, standardized mortality ratios, Poisson probabilities, space and space-time clustering, autocorrelation measures and hierarchical clustering are discussed. Surface techniques mentioned are isolines and trend surfaces.

For map comparisons, Lorenz curves, coefficients of areal correspondence and correlation coefficients have been used. Case-control matching, acquaintance networks, multidimensional scaling and cluster analysis are examples of methods that are based on relative or non-metric spaces. The review gives rise to the discussion of several general points: problems encountered in spatial analysis, theory building and verification, the appropriate role of technique and computer use. Some suggestions are made for further use of spatial analytic techniques in medical geography: Monte Carlo simulation of point patterns, network analysis to study referral systems and health care for pastoralists, GEOGRAPHIC INFORMATION SYSTEMS to assess environmental risk, difference mapping for disease and risk factor map comparisons and multidimensional scaling to measure social distance.

Jacoby, I. 1991. Geographic distribution of physician manpower: the GMENAC (Graduate Medical Education National Advisory Committee) legacy. *Journal of Rural Health*, 7(4 Suppl), 427–436.

The Graduate Medical Education National Advisory Committee (GMENAC) projected the need for and supply of physicians and other providers, recommended time and access standards for health care services, and developed guidelines for the geographic distribution of physicians. Since this study, analysts have given scant attention to national problems of physician geographic distribution. The issue deserves additional scrutiny in light of the current continuing problems of underservice in rural areas. The emergence of

GEOGRAPHIC INFORMATION SYSTEMS offers a unique opportunity to acquire data on provider distribution and provide a framework for developing and testing redistribution policy.

Kitron, U. et al. 1994. Geographic information system in malaria surveillance: mosquito breeding and imported cases in Israel, 1992. *American Journal of Tropical Medicine & Hygiene,* 50(5), 550–556.

Although a significant resurgence of malaria in Israel is unlikely at present, the risk for a localized outbreak of malaria cases due to infection of local anopheline mosquitoes by imported cases does exist. A national computerized surveillance system of breeding sites of Anopheles mosquitoes and imported malaria cases was established in 1992 using a geographic information system (GIS). Distances between population centers and breeding sites were calculated, and maps associating epidemiologic and entomologic data were generated. Risk of malaria transmission was assessed with consideration of vectorial capacity and flight range of each Anopheles species.

The GIS-based surveillance system ensures that if a localized outbreak does occur, it will be associated rapidly with a likely breeding site, a specific Anopheles vector, and a probable human source, so that prompt control measures can be most efficiently targeted. This cost-effective GIS-based surveillance system can be expanded and adapted for countries with indigenous malaria transmission.

Lam, N.S. 1986. Geographical patterns of cancer mortality in China. *Social Science & Medicine,* 23(3), 241–247.

This research note discusses the China cancer mortality data and the methodological problems involved in spatial analysisof these data. Some of the research findings produced by mapping and analyses of the cancer data at the provincial level are also summarized. The two most common cancers in China, stomach and oesophagus, were found to have no significant correlation with some selected physical variables and population density, suggesting the need to examine other socioeconomic variables such as dietary habit.

The study also suggests that the type of diet which may be responsible for these two cancers could be very different from each others. Colon and rectum, leukemia, and breast cancers were found to have very high positive spatial autocorrelation and high correlation with population density — a result contrary to previous findings in the West. Future research using a geographic information system approach and county data is suggested.

Malone, J.B. et al. 1992. Use of LANDSAT MSS imagery and soil type in a geographic information system to assess site-specific risk of fascioliasis on Red River Basin farms in Louisiana. *Annals of the New York Academy of Sciences*, 653, 389–397.

A geographic information system (GIS) was constructed in an ERDAS environment using maps of soil types from the USDA Soil Conservation Service, LANDSAT satellite multispectral scanner data (MSS), boundaries for 25 study farms, and slope and hydrologic features shown in a two-quadrangle (USGS, 7.5') area in the Red River Basin near Alexandria, Louisiana. Fecal sedimentation examinations were done in the fall of 1989, spring of 1990, and fall and winter of 1990–1991 on 10–16 random samples per herd. Fecal egg shedding rates for F. hepatica ranged from 10–100% prevalence and 0.3–21.7 eggs per two grams of feces (EP2G). For *Paramphistomum spp.*, a rumen fluke also transmitted by F. bulimoides but not affected by flukicides, egg shedding rates ranged from 10–91% prevalence and 0.1–42.8 EP2G.

Soil types present ranged from sandy loams to hydric, occasionally flooded clays. Herd *Paramphistomum spp.* egg shedding rates increased with the proportion of hydric clays present, adjusted for slope and major hydrologic features. *F. hepatica* infection intensity followed a similar trend, but were complicated by differing treatment practices. Results suggest that earth observation satellite data and soil maps can be used, with an existing climate forecast based on the Thornthwaite water budget, to develop a second generation model that accounts for both regional climate variation and site-specific differences in fascioliasis risk based on soils prone to snail habitat.

Richards, F.O. Jr. 1993. Use of geographic information systems in control programs for onchocerciasis in Guatemala. *Bulletin of the Pan American Health Organization*, 27(1), 52–55.

(No abstract)

Rogers, D.J. and Williams, B.G. 1993. Monitoring trypanosomiasis in space and time. *Parasitology*, 106 Suppl. pp. S77–92.

The paper examines the possible contributions to be made by Geographic Information Systems (GIS) to studies on human and animal trypanosomiasis in Africa. The epidemiological characteristics of trypanosomiasis are reviewed in the light of the formula for the basic reproductive rate or number of vector-borne diseases. The paper then describes how important biological characteristics of the vectors of trypanosomiasis in West Africa may be monitored using data from the NOAA series of meteorological satellites. This will lead to an understanding of the spatial distribution of both vectors and disease. An alternative, statistical approach to understanding the spatial distribution of tsetse, based on linear discriminant analysis, is illustrated with the example of Glossina morsitans in Zimbabwe, Kenya and Tanzania.

In the case of Zimbabwe, a single climatic variable, the maximum of the mean monthly temperature, correctly predicts the prerinderpest distribution of tsetse over 82% of the country; additional climatic and vegetation variables do not improve considerably on this figure. In the cases of Kenya and Tanzania, however, another variable, the maximum of the mean monthly Normalized Difference Vegetation Index, is the single most important variable, giving correct predictions over 69% of the area; the other climatic and vegetation variables improve this to 82% overall. Such statistical analyses can guide field work towards the correct biological interpretation of the distributional limits of vectors and may also be used to make predictions about the impact of global change on vector ranges.

Examples are given of the areas of Zimbabwe which would become climatically suitable for tsetse given mean temperature increases of 1, 2 and 3 degrees Centigrade. Five possible causes for sleeping sickness outbreaks are given, illustrated by the analysis of field data or from the output of mathematical models. One cause is abiotic (variation in rainfall), three are biotic (variation in vectorial potential, host immunity, or parasite virulence) and one is historical (the impact of explorers, colonizers and dictators). The implications for disease monitoring, in order to anticipate sleeping sickness outbreaks, are briefly discussed.

It is concluded that present data are inadequate to distinguish between these hypotheses. The idea that sleeping sickness outbreaks are periodic (i.e., cyclical) is only barely supported by hard data. Hence it is even difficult to conclude whether the major cause of sleeping sickness outbreaks is biotic (which, in model situations, tends to produce cyclical epidemics) or abiotic. (Abstract truncated).

Sanson, R.L., Pfeiffer, D.U. and Morris, R.S. 1991 Geographic information systems: their application in animal disease control. *Review of Science and Technology*, 10(1), 179–195.

Geographic information systems (GIS) are computerised information systems that allow for the capture, storage, manipulation, analysis, display, and reporting of geographically referenced data. They have been used in recent years for a wide variety of purposes, including town planning and environmental resource management. The technology has many features which make it ideal for use in animal disease control, including the ability to store information relating to demographic and causal factors and disease incidence on a geographical background, and a variety of spatial analysis functions. A number of possible veterinary applications are suggested, and three examples of the use of GIS in New Zealand are discussed.

Stallones, L., Nuckols, J.R. and Berry, J.K. 1992 Surveillance around hazardous waste sites: geographic information systems and reproductive outcomes. *Environmental Research,* 59(1), 81–92.

The purpose of this paper is to discuss the potential for integrating surveillance techniques in reproductive epidemiology with geographic information system technology to identify populations at risk around hazardous waste sites. Environmental epidemiologic studies have had problems with estimating or measuring exposures to individuals, and of detecting effects when the exposure is low, but continuous. In addition, exposures around hazardous waste sites are complex and frequently involve chemical mixtures.

The birth weight of human babies has been reported to be sensitive to many environmental influences. Birth weight can be analyzed as a continuous variable or as a dichotomous one using the standard cutpoint of 2500g or less to indicate low birth weight. It has the potential to be a powerful surveillance tool since exposures to the fetus reflect maternal and paternal exposures. The advent of recent environmental regulations pertaining to hazardous waste sites has greatly increased the availability of environmental data for many sites. The major problem with incorporating these data into epidemiologic studies has been with the logistics of data management and analysis. Computer-assisted geographic information systems hold promise in providing capabilities needed to address the data management and analysis requirements for effective epidemiologic studies around to hazardous waste sites.

Walter, S.D. 1993 Visual and statistical assessment of spatial clustering in mapped data. *Statistics in Medicine,* 12(14), 1275–1291.

Maps have seen increasing use to examine regional variation in health, but there has been little research on the visual perception of spatial patterns in mapped data. Theories of graphical perception suggest that the interpretation of maps is complex relative to other types of graphical material. This paper describes an experiment in which observers assessed a series of maps with respect to their amount of clustering. Maps with various types of spatial pattern were visually distinguishable; comparisons between variants of the same map, however, using different shading and plotting symbols indicated that the method of data representation also had a strong effect on visual perception. There was some evidence for a learning effect in complex maps. The relationship between the visual assessments and a statistical measure of spatial autocorrelation was significant but imperfect.

Wartenberg, D. 1992. Screening for lead exposure using a geographic information system. *Environmental Research*, 59(2), 310–317.

Screening programs for lead overexposure typically target high-risk populations by identifying regions with common risk markers (older housing, poverty, etc.). While more useful than untargeted screening programs, targeted programs are limited by the geographic resolution of the risk-factor information. A geographic information system can make screening programs more effective and more cost-efficient by mapping cases of overexposure, identifying high-incidence neighbourhoods warranting screening, and validating risk-factor-based prediction rules.

Washino, R.K. and Wood, B.L. 1994 Application of remote sensing to arthropod vector surveillance and control. *American Journal of Tropical Medicine & Hygiene*, 50(6 Suppl), 134–144.

A need exists to further develop new technologies, such as remote sensing and geographic information systems analysis, for estimating arthropod vector abundance in aquatic habitats and predicting adult vector population outbreaks. A brief overview of remote sensing technology in vector surveillance and control is presented, and suggestions are made on future research opportunities in light of current and proposed remote sensing systems.

GIS Books, Journals, and Magazines

Books

Antenucci J. et al. 1991. Geographic information systems: A guide to the technology, Van Nostrand Reinhold. ISBN 0-442-00756-6.

Aronoff S., 1989, 1991. Geographic information systems: A management perspective, WDL Publications. ISBN 0-921804-91-1.

(Note: An excellent introduction to the technology. I use this as the textbook for the introductory GIS course I teach at undergraduate geography level.)

Belward A.S., C.R. Valanzuela (eds.), 1991. Remote sensing and geographical information systems for resource management in developing countries, Kluwer Academic Press. ISBN 0-7923-1268-6.

Berry J.K., 1993. Beyond mapping: Concepts, algorithms, and issues in GIS, GIS World. ISBN 0-9625063-2-6.

Burrough, P.A. 1986. Principles of geographical information systems for land resources assessment, Oxford University Press. ISBN 0-89291-175-1.

(A more advanced book which I have used as a text in the second GIS course I teach at undergraduate level. This book is more about GIS principles than land resource assessment so do not be dissuaded by the title.)

Fotheringham S.;. Rogers, P. eds. 1994. Spatial analysis and GIS, Taylor and Francis. ISBN 0-7484-0104-0.

Goodchild M.; Parks, B.; Steyaert. L. eds. 1993. Environmental modelling with GIS, Oxford University Press. ISBN 0-19-508007-6.

Goodchild M.; Gopal. S. eds. 1989. The accuracy of spatial databases, Taylor and Francis. ISBN 0-85066-847-6.

(Some chapters are very technical but this book is an excellent compilation of the issues surrounding error in GIS.)

Huxhold W.E., 1991. An introduction to urban geographic information systems, Oxford University Press. ISBN 0-19-506533-4.

(Another good introduction to the technology but slanted towards urban applications.)

Laurini R.; Thompson, D. 1992. Fundamentals of spatial information systems, Academic Press. ISBN 0-12-438380-7.

(Coverage of technical software issues is wide-ranging but is rather awkward reading at first. Recommend reading another introductory text before reading this one.)

Maguire D.J.; Goodchild, M.; Rhind, D. eds. 1991. Geographical information systems: Principles and applications, Longman Scientific & Technical, (two volumes). ISBN 0-582-05661-6.

(Perhaps the most extensive and definitive GIS book.)

Rhind D.; Raper, J.; Mounsey, H. eds. 1990, 1992. Understanding geographic information systems, Taylor and Francis. ISBN 0-85066-774-5.

Star J.; Estes, 1989, 1990. Geographic information systems: An introduction, Prentice Hall. ISBN 0-13-351123-5.

(As a first book to read on GIS, this is adequate but I prefer Aronoff.)

Tomlin, D. 1990. Geographic information systems and cartographic modeling, Prentice Hall. ISBN 0-13-350927-3.

(Another adequate first read on GIS.)

Journals and Magazines

Cartography and Geographic Information Systems, since 1973 (formerly *The American Cartographer)*, quarterly, American Congress on Surveying and Mapping. ISSN1050-9844.

Computers and Geosciences: An International Journal, since 1974, 8 issues per year, Pergamon Press. ISSN 0098-3004.

*Geographical Systems: The International Journal of Geographical In*formation, Analysis, Theory, and Decision, since 1994, Gordon and Breach Science Publishers.

GeoInfo Systems, since 1991, 10 issues per year, Aster Publishing Co.

Geomatica (former CISM/ACSGS Journal), since 1945, quarterly, Canadian Institute of Geomatics. ISSN 1195-1036.

GIM: International Journal for Surveying, Mapping, and Applied GIS, since 1987, bimonthly, GITC. ISSN 0928-1436.

GIS Asia/Pacific, since 1985, bimonthly, GIS World Inc.

GIS Europe: Europe's Geographic Information Systems Magazine, since 1992, 10 issues per year, Longman Group (Longman Geoinformation). ISSN 0926-3403.

GIS World, since 1988, 10 issues per year, GIS World Inc. ISSN 0897-5507.

International Journal of Geographical Information Systems, since 1986, bimonthly, Taylor and Francis. ISSN 0143-1161.

Journal of Urban and Regional Information Systems, Urban and Regional Information Systems Association.

Mapping Awareness (former Mapping Awareness and Integrated Spatial Information Systems: in the United Kingdom and Ireland), since 1987, 10 issues per year, Longman Group Inc. ISSN 0954-7126.

Surveying and Land Information Systems, since 1940, quarterly, American Congress on Surveying and Mapping. ISSN 0039-6273.

SW: Journal for Land Survey, Hydrographic Survey and Land Information Management, since 1993, bimonthly, GITC, ISSN 0-927-7900.

GIS Software

This is a selected list of GIS software packages with contact information. Given the rapidly changing nature of GIS software in terms of price and functionality, no attempt has been made to describe and/or compare these packages. The information given here is based on the "GIS Package List" maintained by Oliver Weatherbee. Weatherbee's list is more comprehensive than the information given here and includes descriptions of the packages and reviewer comments. It can be obtained through the internet at http://www.laum.uni-hannover.de. The reader is also referred to the vendors directly for their latest information.

Name: ARC/INFO, pcARC/Info, ArcCAD, ArcView
Type: Commercial
Operating Systems: UNIX, DOS and Windows
Company: Environmental Systems Research Institute, Inc. (ESRI)
380 New York Street
Redlands, California 92373
USA
Phone: 909-793-2853
Fax: 909-793-2853

Name: ATLAS-GIS
Type: Commercial
Operating Systems: DOS and Windows
Company: Strategic Mapping, Inc.
Suite 250
4030 Moorpark Ave.
San Jose, California 95117
USA
Phone: 408-985-7400
Fax: 408-985-0859

Name: CARIS
Type: Commercial
Operating Systems: UNIX, Windows
Company: Universal Systems Ltd.
270 Rookwood Avenue
P.O. Box 3391, Station B
Fredericton, New Brunswick,
Canada
Phone: 506-458-8533
Fax: 506-459-3849

Name:	CISIG
Special Note:	A low-cost GIS that is distibuted with the digitizing package
ROOTS.	Lower cost for developing countries (US$200).
Operating Systems:	DOS (?)
Company:	GIS Services Conservation Planning and Technical Cooperation Conservation International 1015 18th Street, NW (Suite 1000) Washington, DC 20036 USA Phone: 202-429-5660 Fax: 202-887-0193

Name:	GENAMAP
Type:	Commercial
Operating Systems:	UNIX
Company:	Genasys II Pty. Ltd. 13th Level, 33 Berry Street, N. Sydney, New South Wales 2060 Australia Phone: 61-2-954-0022 Fax: 61-2-954-9930

Name:	GRASS
Type:	Public Domain (freeware)
Operating Systems:	UNIX (Mac under development)
FTP site:	moon.cecer.army.mil (129.229.20.254) in /grass directory.

Name:	IDRISI
Type:	Commercial (low cost)
Operating Systems:	DOS (Windows version due May, 1995)
Company:	Graduate School of Geography Clark University 950 Main St, Worcester, Massachusetts 01610 USA Phone: 508-793-7526 Fax: 508-793-8842

Name:	ILWIS
Type:	Commercial
Operating Systems:	DOS (?)
Company:	ILWIS
	International Institute for Aerospace Survey and Earth
Sciences	
	350, Boulevard 1945
	P.O.Box 6
	7500 AA Enschede
	The Netherlands
	Phone: 31-53-874337
	Fax: 31-53-874484

Name:	Intergraph MGE
Type:	Commercial
Operating Systems:	UNIX, Macintosh and Windows NT.
Company:	Intergraph Corp.
	Mapping Sciences Division
	Huntsville, Alabama 35894
	USA
	Phone: 205-730-2700

Name:	MAPGRAFIX
Type:	Commercial
Operating Systems:	Macintosh
Company:	ComGrafix, Inc.
	620 E Street
	Clearwater, Florida 34616
	USA
	Phone: 813-443-6807
	Fax: 813-443-7585

Name:	MAPINFO
Type:	Commercial
Operating Systems:	UNIX, Macintosh and Windows.
Company:	MapInfo Corp. One Global View Troy, New York 12180 USA Phone: 518-285-6000 Fax: 518-285-6060

Name:	MOSS
Type:	Public Domain: PC version via FTP, UNIX version US$500.
Operating Systems:	DOS, UNIX. (DOS version is several revs. behind UNIX and no longer updated)
FTP:	(DOS version only) ftp.csn.org (128.138.213.21) in /cogs/moss
Company:	Bureau of Land Management Service Center Denver Federal Center Denver, Colorado 80225 USA

Name:	OzGIS
Special Note:	This package can be obtained as shareware by FTP for demonstrationpurposes. Actual cost of software is approx. US$500.
Operating Systems:	DOS and Windows.
FTP:	ftp.cica.indiana.edu in /pub/pc/win3/misc.
Company:	The Clever Company QMDD Box 6108 Queanbeyan 2620 Australia

Name:	pMAP	
Type:	Commercial	
Operating Systems:	DOS	
Company:	Spatial Information Systems, Inc.	
	19 Old Town Square	
	Fort Collins, Colorado 80524	
	USA	
	Phone:	303-490-2155
	Fax:	303-482-0251

Name:	SPANS	
Type:	Commercial	
Operating Systems:	UNIX, DOS and OS/2.	
Company:	Intera TYDAC Technologies Inc.	
	2 Gurdwara Road, Suite 210	
	Nepean, Ontario K2E 1A2	
	Canada	
	Phone:	613-226-5525
	Fax:	613-226-3819

Name:	WINGIS	
Type:	Commercial	
Operating Systems:	Windows/Windows NT	
Company:	PROGIS w.h.m. G.m.b.H.	
	Italienerstr. 3, A-9500,	
	Villach,	
	Austria	
	Phone:	42-42-26332

Disclaimer

This author, Steven Reader, accepts no responsibility for any errors in this list, nor endorses any of the products listed above in any way.

GIS Internet Resources

As with many fields, the global computer network known as the Internet is rapidly becoming a major resource for GIS information, data, software and expertise. The list of sites below is a selection intended to provide starting points for exploring GIS resources on the Internet. These resources are growing daily and should be monitored periodically. No attempt is made here to explain the different modes of access of these resources. The reader is referred to any introductory book on the Internet for this information.

Useful World Wide Web (WWW) Servers

All these sites are useful starting points for GIS resource exploration on the WWW.

http://www.geo.ed.ac.uk	University of Edinburgh, Scotland.
http://info.er.usgs.gov	United States Geological Survey.
http://www.laum.uni-hannover.de	University of Hannover, Germany
http://www.tis.psu.edu	Penn State University, USA
http://www.iko.unit.no	Norwegian Institute of Technology
http://deathstar.rutgers.edu	Rutgers University, USA
http://www.ncgia.ucsb.edu	National Center for Geographic Information and Analysis, USA

All the above sites contain pointers to other GIS-oriented sites. To get a fairly comprehensive list of GIS-oriented WWW home pages, use *anonymous ftp* to retrieve the following:

ftp://gis.queensu.ca/pub/gis/docs/gissites.txt	Michael McDermott's list

A GIS frequently-asked-questions (FAQ) is maintained at the following location:

http://www.census.gov	United States Bureau of the Census

and through anonymous ftp from abraxas.adelphi.edu in /pub/gis/FAQ.

International Workshop on Geographic Information Systems (GIS) for Health and the Environment

Colombo, Sri Lanka
5–10 September 1994

AGENDA

Monday, 5 September

9:00–10:00 Opening Ceremony
Chairperson: **Kamini Mendis**

Panduka Wijeyaratne, Environmental Health Project, USAID, USA
Srimani Athulathmudali, Hon. Minister for Transport, Environment and Women's Affairs, Sri Lanka
N. Kodagoda, Vice Chancellor, University of Colombo
Gilles Forget, Health Sciences Division, Canada
W.P. Fernando, Anti-Malaria Campaign, Ministry of Health, Sri Lanka
A.R. Wickremasinghe, Department of Community Medicine, University of Peradeniya and University of Colombo, Sri Lanka

10:00–10:30 Tea break

10:30–10:45	*Overview of Agenda and Objectives of Workshop* **Don de Savigny**, Canada **A.R. Wickremasinghe**, Sri Lanka
10:45–12:45	Session 1. *GIS, Health, and the Environment* Chairperson: **Panduka Wijeyaratne** *The State of GIS Technology and Future Trends* **S. Reader** *Health and Disease: A Geographical Approach* **I. Nuttall**
12:45–14:00	Lunch
14:00–15:30	Session 2. *GIS-related Technologies for Monitoring and Control of Disease Introduction* Chairperson: **A.H. Dhanapala** *The Use of Global Positioning Systems (GPS) and GIS in Malaria Research, Management and Control in South Africa* **D. Lesuer** and **S. Ngxongo** *The Use of Low Cost Remote Sensing and GIS in Identifying and Monitoring the Environmental Factors Associated with Vector-Borne Disease Transmission* **S.J. Connor** *Diagnostic Featuress of Malaria Transmission in Nadlad (Gujarat) Using Remote Sensing and GIS* **A. Srivastava** and **M.S. Malhotra** *Global Data Needs for GIS and Tropical Disease Control* **P. Herath**
15:30–15:50	Tea break
15:50–17:00	*GIS Needs Assessment* Participants will meet with resource persons and discuss their GIS project needs and problems

Tuesday, 6 September

8:30–10:00 Session 3. *GIS and the Monitoring and Management of Environmental Toxicants*
Chairperson: **G. Forget**

The Use of the RAISON System for Water Quality Data Management in Malaysia
C.W. Wang

A New Biological Method to Monitor the Toxicant Pollution in the Aquatic Environment
Y. Shen, M. Gu and W. Feng

Protecting the Competitiveness of Mining Products: Development of Guidelines for the Regulations of Arsenic Concentration in the Environment
C. Ferreccio and N.M. Pareto

L'expérience du Sénégal de Entegration des données envlronmentales et sanitaires pour la région de Fleuve
L.G. Sarr and A.H. Sylla

10:00–10:15 Coffee break

10:15–11:00 *Summary of Issues from Previous Sessions (1-3)*
S. Reader and F. Nobre

11:00–12:30 Session 4. *GIS in Health Research I: Malaria*
Chairperson: **Don de Savigny**

Geographical Information Systems in the Study and Control of Malaria
G. Brêtas

The Application of GIS to Evaluate the Quality of Indoor Spraying in Malaria Control in Botswana
D.W. Rumisha

	Geographic Information Systems in Malaria Surveillance in Israel **U. Shalom**
12:30–14:00	Lunch
14:00–15:00	Session 5. *GIS Health Research II: Malaria in Sri Lanka* Chairperson: **I. Nuttall**
	The Use of GIS in the Study of FActors Influencing Malaria Transmission in an Endemic Administrative District of Sri Lanka **R. Wickremasinghe, T. Abeysekera, et al.**
	Spatiala Analysis of Malaria Risk in an Endemic Region of Sri Lanka **D.M. Gunawardena, L. Muthuwatta, et al.**
15:00–15:15	Tea break
15:15–16:45	Session 6. *GIS in Health Research III* Chairperson: **L. Loslier**
	The Utilization of GIS Techniques in the Control of Schistosomiasis in Botswana **T. Pilatwe**
	A. GIS Approach to Determination of Catchment Populations Around Local Health Facilities in Developing Countries **H.M. Oranga**
	Dengue Hemorragic Fever in the Kingdom of Cambodia **T. Davuth**
	GIS in Research and Control of Leishmaniasis in Tunisia **R. Ben-Ismail**
16:45–17:30	*Summary of Issues from Previous Sessions (4–6)* **I. Nuttall and K. Kotta**

Wednesday, 7 September

7:00	*Field Visit to Malaria Study Area*

Thursday, 8 September

9:30–17:30	*GIS hands on*

Friday, 9 September

9:00–17:30	*GIS hands on*

Saturday, 10 September

9:00–11:00	Session 7. *Panel Discussion — Issues and Analyses in Terms of Workshop Objectives* Discussants: R. Ben-Ismail, L. Loslier, A. Srivastava, F. Binka, C.W. Wang Resource persons and participants will summarize their views and present follow-up plans, including networking.
11:00–11:15	Tea break
11:15–12:30	*Follow-Up* **Don de Savigny** *Closing Remarks* G. Forget, IDRC M.M. Ismail, University of Colombo K.N. Mendis, University of Colombo A.H. Dhanapala, University of Colombo

Geographic Information Systems for Health and the Environment

Participants

Dr Thisula Abeysekara
Malaria Research Unit
Department of Parasitology
Faculty of Medicine
P.O. Box 271, Kynsey Road
Colombo 8, Sri Lanka

Dr Felix Amerasinghe
Department of Zoology
Faculty of Science
University of Peradeniya
Peradeniya, Sri Lanka
Tel: 08 88693 ext. 204
Fax: 08 88018
e-mail: FELIX@SCI.PDN.AC.LK

Mr Senerath Bandara
Regional Malaria Officer
Anti-Malaria Campaign
Moneragala District
Ministry of Health Services
Moneragala, Sri Lanka

Dr Riadh Ben-Ismael
Institut Pasteur de Tunis
Laboratorie d'epidemiologic et d'eologie Parasitaire
13 Place Pasteur
B.P. 74, Tunis-Belvedere 1002
Tunisie
Tel: (216) 1 792 429
Fax: (216) 1-791 833

Dr Fred Binka
Epidemiologist
Navrongo Health Research Centre
Ministry of Health
P.O. Box 114
Navrongo, Ghana
Tel: (233) 723 425
Fax: (233) 072-3425
e-mail: FBINKA@GHA2.HELATHNET.ORG

Dr Gustavo Bretas
London School of Tropical Medicine
Keppel Street
London WC1E 7HT
United Kingdom
Tel: (71) 9272053/6270758
e-mail: GBRETAS@LSHTM.AC.UK

Dr Stephen Connor
Liverpool School of Tropical Medicine
Pembroke Place
Liverpool L3 5QA
United Kingdom
Tel: (51) 708-9393
Fax: (51) 708-8733
e-mail: c/o.MSERIVCE@LIVERP.O.OL.AC.UK

Dr Tiv Davuth
Department of Biochemistry
Faculty of Medicine
National Centre for Hygiene & Epidemiology
226 Street Kampuchea Krom
Khan 7 - Makara
Phnom Penh, Cambodia
Tel: (855) 236-6205
Fax: (855) 232-6841

Dr Don de Savigny
Senior Program Specialist
International Development Research Centre
Health Sciences Division
250 Albert Street
Ottawa, ON Canada K1G 3H9
Tel: (613) 236-6163
Fax: (613) 567-7748
e-mail: DDESAVIGNY@IDRC.CA

Dr Stanley de Silva
Deputy Director
Anti-Malaria Campaign
555/5 Elvitigala Mawatha
Colombo 5, Sri Lanka

Dr A.H. Dhanapala
Senior Lecturer
Department of Geography
University of Colombo
Kumaratunga Munidasa Mawatha
Colombo 3, Sri Lanka

Dr F.C. Ferreccio
Monjitas 454, Of. 109
Casilla 52750
Correo Central
Santiago, Chile
Tel: (56) 2 638-4148
Fax: (56) 2 632-7256

Dr Gilles Forget
Director
International Development Research Centre
Health Sciences Division
250 Albert Street
Ottawa, ON Canada K1G 3H9
Tel: (613) 236-6163
Fax: (613) 567-7748
e-mail: GFORGET@IDRC.CA

Dr Renu Goonewardena
Malaria Research Unit
Department of Parasitology
Faculty of Medicine
P.O. Box 271, Kynsey Road
Colombo 8, Sri Lanka
e-mail: DRG@PARASIT.CMB.AC.LK.

Mr D.M. Gunawardene
Malaria Resesarch Unit
Department of Parasitology
Faculty of Medicine
P.O. Box 271, Kynsey Road
Colombo 8, Sri Lanka
Fax: (94) 1-699 284 (fax/phone)

Dr Pushpa Herath
Control of Tropical Diseases
World Health Organization
CH - 1211 Geneva 27
Switzerland
Tel: (41) 22 791-3645
Fax: (41) 22 791-0746
e-mail: HERATH@WHO.CH

Dr Frank Indome
Navrongo Health Research Centre
Ministry of Health
P.O. Box 1124
Navrongo, Ghana
Tel: (233) 723 425
e-mail: FINDOME@GHA2.HEALTHNET.ORG

Mr Gamini Jayasundara
Malaria Research Unit
University of Colombo
Department of Parasitology
Faculty of Medicine
University of Colombo
P.O. Box 271, Kynsey Road
Colombo 8, Sri Lanka

Dr Shaden Kamhawi
Dept. of Biology, Faculty of Sciences
Yarmouk University
Irbid, Jordan
962-6-614145 or 640221
FAX (at home) 962-6-662494

Mr H.M. Kodisinghe
Regional Malaria Officer
Anti-Malaria Campaign
Kurunegalla
Department of Parasitology
University of Colombo
P.O. Box 271, Kynsey Road
Colombo 8, Sri Lanka

Mr Flemming Konradson
Health and Irrigation Program
International Irrigation Management Institute
P.O. Box 2075
Colombo 8, Sir Lanka

Mr P.K. Kotta
GIS Consultant
South Asian Cooperative Environment Program
84 Lorensz Road
Colombo 4, Sri Lanka
Tel: (941) 582 533
Fax: (941) 589-369

Mr Wasantha Udaya Kumara
Malaria Research Unit
Department of Parasitology
Faculty of Medicine
P.O. Box 271, Kynsey Road
Colombo 8, Sri Lanka

Dr Davis LeSueur
Sepcial Scientist
Medical Research Council
P.O. Box 17120
Congella 4013, South Africa
Tel: (31) 251 481
Fax: (31) 258 840
e-mail: LESUEUR@MED.UND.AC.ZA

Prof. Luc Loslier
Department of Geography
University of Quebec
P.O. Box 8888, Station A
Montreal, PQ H3C 3P8
Canada
Tel: (514) 987-3682
Fax: (514) 987-3682
e-mail: LOSLIER.LUC@UQAM.CA

Dr M.S. Malhotra
Senior Research Officer
Malaria Research Centre
22 Sham Nath Marg
Delhi 110054, India
Tel: (11) 224-7983/224-3006
Fax: (11) 221-5086

Prof. Kamini Mendis
Malaria Research Unit
Department of Parasitology
Faculty of Medicine
University of Colombo
P.O. Box 271, Kynsey Road
Colombo 8, Sri Lanka
Fax: (941) 699 284 (fax/phone)

Mr Lal Muthuwatte
Research Assistant
Department of Engineering Mathematics
Open University of Sri Lanka
Nawalga, Nugedoda
Sri Lanka
e-mail: LPM@MATH.AC.OU.LK.

Dr Niroshini Nirmalan
Lecturer
Department of Parasitology
Faculty of Medicine
University of Colombo
P.O. Box 271, Kynsey Road
Colombo 8, Sri Lanka
Tel: (941) 688 660

Dr Isabelle Nuttall
Division of Control of Tropical Diseases (TDR)
World Health Organization
CH-1211 Geneva 27
Switzerland
Tel: (41) 22 791-3861/791-3898
Fax: (41) 22 791-0746
e-mail: NUTTALL@WHO.CH

Dr Sipho M. Ngxongo
Control Health Inspector
Department of Health
P.O. Box 002
Joun 3969
South Africa
Tel: (27) 035672-11
Fax: (27) 035672-21

Dr Flavio Fonseca Nobre
COPPE/UFRJ
Programa de Engenharia Biomedica
Universidade Federal de Rio de Janeiro
Caixa Postal 68510
21945-970 Rio de Janeiro, Brazil
Tel: (55) 21 230-5108
Fax: (55) 21 290-6626
e-mail: FLAVIO@SERV.PEB.UFRJ.BR

Dr Hezron Oranga
Department of Health Policy and Management
African Medical and Research Foundation
P.O. Box 30125
Nairobi, Kenya
Tel: (254) 2 506 112
Fax: (254) 2 501 301/302/501 331

Dr T. Pilatwe
Statistician
Ministry of Health
Private Bag 00269
Gaborone, Botswana
Tel: (267) 374 353, ext. 229
Fax: (267) 374 351

Dr D.W. Runisha
Epidemiologist
Community Health Services
Private Bag 00269
Gaborone, Botswana
Tel: (267) 374 354
Fax: (267) 374 351

Dr Steven Reader
Deptartment of Geography
McMaster University
1280 Main Street West
Hamilton, ON L8S 4K1
Canada
Tel: (905) 525-9140
Fax: (905) 546-0463
e-mail: READER@MCMASTER.CA

Dr (Colonel) Lamine Cisse Sarr
Directeur de la Santé Publique
Ministère de la Santé
Dakar, Senegal
FAX: 221-238-488

Dr Uri Shalom
Head
Department of Pest Surveillance & Control
Ministry of Environment
Jerusalem 91061, Israel
Tel: (972) 2 231 407/151 853
Fax: (972) 2 251 830

Dr Soumbey-Alley
WHO/OCP
P.O. Box 549
Ouagadougou, Burkina Faso
FAX: 226-30-21 47

Dr (Mrs) Aruna Srivastava
Senior Research Officer
Malaria Research Centre
22 Sham Nath Marg
Delhi 110054
India
Tel: (91) 11 224-7983/224-3006
Fax: (91) 11 221-5086

Dr Annette Stark
International Development Resesarch Centre
P.O. Box 101, Tanglin 9124, Singapore
Tel: (65) 235-1344
Fax: (65) 235-1849
e-mail: ASTARK@IDRC.CA

Mr A.H. Sylla
Statisticien
Ministère de la Santé
P.O. Box 4024/4006
Dakar, Senegal
Tel: (221) 247 549
Fax: (221) 238-488

Dr Idrissa Talla
11 Endsleigh Gardens, Flat 3
London WC1
United Kingdom

Dr C.W. Wang
Dept. of Biochemistry
Faculty of Medicine
University of Malaysia
59100 Kuala Lumpur
Malaysia
Tel: (603) 750-2948
Fax: (603) 755-7740/756-8841

Dr Rajitha Wickremasinghe
Deptartment of Community Medicine
Faculty of Medicine
University of Peradeniya
Peradeniya, Sri Lanka
Tel: (941) 699 284/(948) 88130
Fax: (941) 699 284/(948) 32572
e-mail: RAJWICKS@PARASIT.CMB.AC.LK.

Dr Panduka Wijeyaratne
Program Director - Tropical Diseases
Environmental Health Project (EHP)
C/O CDM
1611 North Kent Street
Arlington, VA 22209-2111
USA
Tel: (703) 247-8763/247-8730
Fax: (703) 243-9004

Prof. Shen Yun-Fen
Institute of Hydrobiology
Academia Sinica
Wuhan, Hubei 430072
China
Tel: (86) 27 782-3481/781-2180
Fax: (86) 27 782-5132